Also by Howard Kohn:
WHO KILLED KAREN SILKWOOD?

THE LAST FARMER

An American Memoir

by Howard Kohn

Summit Books

New York London Toronto Sydney Tokyo

Simon & Schuster Building
Rockefeller Center
1230 Avenue of the Americas
New York, New York 10020

Published by SUMMIT BOOKS
SUMMIT BOOKS and colophon are trademarks of Simon & Schuster Inc.
Designed by Levavi & Levavi
Manufactured in the United States of America

10 9 8 7 6 5 4 3 2 1

Library of Congress Cataloging in Publication Data

Kohn, Howard.
The last farmer.

1. Kohn, Frederick. 2. Kohn, Howard.
3. Farmers—Michigan—Biography. I. Title.
S417.K627K64 1988 977.4'47 [B] 88-8629

ISBN 0-671-49803-7

For Fredrick and Clara Kohn,
my father and mother

ACKNOWLEDGMENTS

I want to thank the following for their contributions to this book: Dorothy Stieve, who helped with the historical research; Dominick Anfuso, Arthur Samuelson, David Bain, Ron Powers, Paul Hendrickson, Ron Kroese, David Weir and Loretta Fidel, who advised and encouraged my writing; Kathy Robbins and Jim Silberman, who had faith in me; Jann Wenner, who first convinced me to write about my father's farm; and especially, for their understanding, all of those whose lives are on these pages.

PROLOGUE

Halfway up the Lake Huron coast of Michigan is the Saginaw Valley, dug and silted by glaciers, and near the northern extreme of the valley, in a Germanic community, is the farm that Heinrich Kohn homesteaded and passed on to his son, Johann Kohn, who passed it to his son, Fredrick Kohn, who is my father. The farm did not pass on to me. When I was eighteen my time on the farm came to an end, that slow, unforgettable time of cow-milking and haying and woodcutting, chores upon chores in raw weather, an unchanging time, a time that seemed a sentence of indeterminate length from which escape at times was doubtful. I left in August of 1965 to attend the University of Michigan, and from there I moved on to addresses in California, New York, Florida, Washington, D.C., and elsewhere. I felt, when I left, that I had been granted a new life: the traveling life of a writer, as it turned out.

Sixteen years later I happened to be back at the farm on a stopover between magazine assignments. Diana, my second wife, was with me. A light rain fell before dawn on the third day of our visit, August 29, 1981, but when it stopped, I expected my father to have us out in the navy beans, sickling the last of the summer weeds. This was a weekday, a workday, and I certainly did not expect him to say, as he did, "Go ahead, go enjoy yourselves." Six of us—Diana, myself, my sister, her husband, and one of my brothers and his wife—set off for the Tittabawassee, the closest navigable river to the farm, looking for a place to rent canoes. I had canoed on rivers around the country, although never on the Tittabawassee. In the last century the Tittabawassee moved the timber of the forested valley to sawmills, and in this century it carries off discharges from the factories of Dow Chemical Company, which has its world headquarters nearby in Midland. I did not associate the Tittabawassee with sport. But we found a canoe livery and plunked three canoes in the river and were miles out of reach when meanwhile, on the farm, my father fell off the roof of the farmhouse.

He had monkeyed onto the wet, steeply pitched roof to try to open a window that could not be budged from the inside of the house. Straining for leverage, he slipped. Several of his grandchildren saw him fall. He landed on his lower back, half on the sidewalk, half on the lawn. In that moment, after almost seventy years of the farming life without one night in a hospital (or one day in debt, or, as far as I knew, without any regrets), after holding fast to principle, and after much, much more, having been brought down suddenly to earth, flattened on his back, wind knocked out, sweat drying on his skin, an elbow opened to the bone, three vertebrae cracked, gazed at by granchildren fearing the worst—how mortal are these prideful old Germans, after all!—at that most panicky and reducing of moments, my father, against all common sense, picked himself up and went about his day as if nothing had happened.

Had he died that morning, I cannot say how the past few years would have turned out. He and I, and whatever lay between us, would have been fixed at that point—he in his place in Old America, with his Germanic standards on his one hundred and twenty-acre farm in Beaver Township, the world of his father and his grandfather; and I in no particular place at all.

But he did not die. He forced himself erect and walked into the house. He let my mother wash and wrap his elbow. Then, as planned, he got into his pickup to give a ride to my brother Roy's wife, Lorie, who needed to go to Detroit to retrieve Roy's pickup, abandoned there by joyriders after a long, illegal ride.

Lorie saw my father's pain, a monumental pain. "We can wait and do this tomorrow," Lorie said, hoping her tone gave the least possible offense. A moment went by with only a shake of his head, and, familiar with his obstinacy, she accepted that the trip would be today. Face toward the road, eyes shifting, she sat next to my father in his truck. Onto Beaver Road, then U.S. 23, they picked up speed. How quickly could she grab the wheel if he went blank? she wondered. Only a few months before he had suffered the first seizure of a weakening heart. At a U.S. 23 exit, pulling off for gas, my father pumped it himself. The detached observer within him knew the pain would not pass and knew that he might heave face first onto greasy asphalt beneath the self-service sign. There was no way he would admit his mistake, even so. He managed to climb back into his pickup. Arriving at the pound for stolen cars, he stayed in his seat, hands braced, while Lorie completed the paperwork. Heading home, Lorie now in Roy's reclaimed pickup, driving a few truck lengths behind my father, she kept a vigil at speeds that were dangerous but nonetheless barely above the minimum, unsuitable for the freeway. Horns sounded. The round trip was two hundred and fifty miles, and altogether they were in traffic seven hours in a slow rain.

While they were still on the road, Diana and I and the others finished our canoe trip. In front of the farmhouse the grandchildren were strangely listless. They were not playing. They ran to us and announced how Grandpa had fallen to the ground. My mother was in the kitchen, busying herself. "The rain, that's why he's late getting back from Detroit," she said to us. She pressed bread dough into loaf pans. "I couldn't stop him from going. Not him."

"I know." I squeezed her shoulders.

The grandchildren began to yell. My father's truck was in the driveway. With tottering baby steps, he worked his way to the front stoop. He clutched the railing to keep his balance and took the long route along the walls to the bedroom. He lay down, still refusing to concede what was evident to everyone and what was repeatedly urged upon him—that he must go to a hospital. "Nope, nope. A little rest, I'll be as good as new." At last my sister Sandra shooed us out of the bedroom and, talking alone with my father, said something that got him on his feet and back into his truck. ("I told him I was going to stand there and yell at him till he gave in, and he knows I'm as stubborn as him.") He let me drive, but he would not let me assist him into the truck or, at the Bay Osteopathic Hospital, out again, pushing aside also, with a veined hand, an orderly who held out a wheelchair. "I don't belong here," my father said foolishly. "Wouldn't be here either if the kids hadn't seen me fall. Wouldn't anybody ever known about it." Intensely nervous, unable to stop looking at him with her red, dry eyes, my mother followed alongside into the emergency room and then to the x-ray lab. She did not ask questions. She was with him not so much to learn the medical facts or be a comfort as to certify that he was still alive. Weeks of hospitalization, months of therapy, even permanent disability she could live with—not to be a widow was all she prayed for.

Blood had drained from his face. I knew this description from books but had not understood it until now. All the brown of a lifetime in the sun was gone. He had the false, bleached color of something not alive. A young nurse, who had eased him finally into a wheelchair, wheeled him from the x-ray lab to a bed. He stood and lifted a leg to get in. "Hold it, stand right there!" a doctor fairly shouted. "I don't want you moving by yourself." The doctor held up a film from the x-ray machine. "It appears you have fifteen fractured vertebrae Mr. Kohn. Please stand very still." He looked at my father, who cuts a man's-man figure. "Okay, we're going to need a male attendant." Fifteen fractures of the bones connecting my father's spine! It was impossible that he could walk! In a council of doctors, there was talk of a cast from neck to waist. Not until a radiologist overruled the initial reading of the x-ray technicians did any of us realize that twelve of the fractures were already healed. They were from old accidents. No one could tell when they had occurred. My father remembered falling onto his back at least twice before, once in the haymow, another time from a wagon, and in each case he had walked off the pain. "I'd have been fine this time, too. Don't know what all the fuss is about," he said, his voice weak now to the point of whispering.

The doctors ordered him into a back brace of steel and canvas, a straitjacket apparatus. They wanted him to have hospital care a minimum of six weeks. You could see them at his bedside, clinical, experienced, familiar with his Germanic streak and dismayed by it. They warned him endlessly of possible paralysis if their every instruction wasn't observed: a standing provocation to a man of spirit. My father removed the brace, and when it was put on again, he loosened it, adjusted it, slipped out of it at night when the nurses weren't vigilant, and, less than a week after his fall, he went home. "You don't use your muscles, they'll turn to noodles," he claimed. He was acting a little, but he believed what he said, talking like a wise old dog. And he had to be aware of the effect he was creating. Happy as my mother was to have him back, she worried every time he did his chores that permanent damage was being done, and I didn't know how long I should stick around to help out. He kept saying he was "good as new," ignoring us in our apprehension, even the Reverend Leroy Westphal, the minister from Beaver Zion Evangelical Lutheran, who had told him good-naturedly at the hospital, "I know it's going to take some tough angels to carry you out of this world, Fred, but why tempt them?"

....

In the war against Hitler, my father was in an Army artillery unit, where he picked up the nickname "von Kohn," a slur that, loyal American, he strived to overcome, although in the rest of his life I believe he would have felt cheated if God had not stamped him out Germanic. The entire time I was growing up I knew him as a man who asserted himself in ways that registered with people.

My father served for a long time on the Beaver Township zoning board, and, at one point, Lyman Schroeder, a neighboring farmer on Flajole Road, sought permission from the board to convert what remained of his farm into a subdivision of trailer homes. Already the board had allowed him to lay out most of it in an eighteen-hole golf course called Sandy Ridge. The Schroeders were one of Beaver's pioneer families, and I knew and liked Lyman. He had hired my brother Ronald and me in our middle teens to help construct the golf course. We planted the matted green plugs, grubbed brush, dug in water pipes, and Lyman gave us a bonus, a lifelong pass to play free. My father had voted in favor of a zoning exemption for the golf course, unhappy though he was at seeing one of the old homesteads pass out of existence, because Lyman was in a fix. Lightning had set fire to his barn and destroyed his dairy herd, and the golf course, underwritten by a federal recreational program, seemed like his last option. Then the state highway crews finished M 20, a new humming connection between Bay City and Midland that brought golfers to within seven miles of Sandy Ridge, turning his luck dramatically around, and so there was far less sympathy when, the second time around, he asked for the trailer-park exemption. Had my father kept quiet, though, it may well have been granted, if only because Lyman's brother Elmer was the township clerk and had considerable influence. Instead, my father raised objections. It would mean giving up more farmland, and for what?—a congestion of people boxed into trailers, people who try on a rind of soil to grow tomatoes up trellises. Their cars would jam the roads, their garbage would overflow, and, almost inevitably, the township would have to have paved thoroughfares and trash collectors and have to get connected to city sewage and water lines. Also, my father had discovered that a development firm was lined up to buy the trailer park: entrepreneurs from outside the valley were behind the scenes. More sprawl was bound to follow. ("I was accused of being the one standing against progress, but, the truth was, most everybody was with me. They just let me do the talking.")

In some communities my father might have been a hardcase, but to his fellow Germanic farmers, speaking the cool, sober, strong-minded language of Beaver Township, not so dissimilar from an upper-class language, he was "somebody who knew his mind." They tended to agree, afterward, that it was his outspokenness—he is not strictly oratorical but when carried away in a debate he has a certain eloquence—and his force of personality, as well as the general regard there is for him, that turned the vote against the trailer park. It was a bitterly charged affair. Somewhat vengefully, Elmer Schroeder dug through minutes of past zoning-board meetings and found that there had been a technical mistake in my father's most recent reappointment to the board. Elmer made it public at the next meeting. My father, in his tenth year as a zoning commissioner, went home and wrote out his resignation on a piece of scrap paper. I cannot say that at the time any of this impressed me, but now, looking back, it is easier to see how honorable he was. He continued to live exactly as before, and things took their course. Elmer died; Lyman grew unhappy and moved to California; and the Schroeder family as a whole, at one time the largest of the pioneer families in the township, became dispersed, its number shrinking from thirty or forty to just two elderly widows.

The rest left, as everyone not content with long, solitary hours and with plainness, and with those few rewards that can be had in Beaver, always leaves. As I left, a few days after turning eighteen. Nothing could have stopped me. Yet leaving was difficult for me, more difficult than I had imagined, and, in the end, I let it deteriorate into a kind of taunt. I got on my high horse. I was a straight-A high school student and bound for college, and I had a summer job on the *Bay City Times* sports desk, but I was also a talked-about jailbird, arrested for drunk and disorderly and for carrying illegal weapons. A disparate rebellion to be sure: hitting the books, hitting the bottle. But I was trying at every turn to put distance between myself and the farm. The elder son on a Germanic farm is the one to carry it through the next generation, and I was the oldest of Fredrick Kohn's sons.

....

There is a recent prototype work in the sociology of American farmers, "Ethnic Communities and the Structure of Agriculture," published in 1985, which rings true. The author is University of Illinois anthropologist Sonya Salamon, and, according to her study, a farmer of Germanic blood is little changed today from grandfathers or great-grandfathers who immigrated with the Middle European wave of 1850–1900 and who, from the first moment off the boat, were "concerned with continuity and tradition and unconcerned if economic progress came slowly." America's Germanic farmers have been known in academic literature for some time as "yeomen." (We are discussing here not the Amish or the Mennonites, two German lines that in America are cultish and statistically marginal, but rather those Germanics surviving on family-scale farms in the heartland, perhaps 400,000 of them, grouped here and there in communities as compact as a township, six miles by six, with one or more churches, either Lutheran or Catholic, as the social center.) On the spectrum of ambition and financial derring-do the Germanics are at the opposite extreme from farmers of Anglo-Saxon descent, "Yankees," for whom "success is measured by criteria from our Protestant ethic, size and profit" and who run their farms "as a business . . . unsentimentally." The Germanics are full of sentiment, for their particular acres and for a family presence on them. "They make continuity an obsessive priority," Dr. Salamon writes, and she quotes an only son, interviewed in 1983, who knows the pressure: "I've been hearing since I was little that I'm expected to carry on the farm." For all of that, Germanic children have been getting out, as fast as we can, you might say. Of children born to Germanic farmers between 1900 and 1980, five of six found other livelihoods, and, to exaggerate only a little, this was as true in 1925 as in 1975, proof of an underlying assumption of America: that the attractions of a better life, higher education, a steady income, free weekends, do draw us from our many ghettos into the mainstream. What has kept the Germanic communities alive have been Germanic mothers, prolific with heirs, from among whom, if a first-born was prodigal, there might be a second-born to carry on, or a third-born, or a son-in-law, or a cousin or nephew. In my mother's case, she had five sons and one daughter. Sandra, Ronald, Harvey and Roy, upon marrying, set up households no more than fifteen minutes from the farm and stayed in good standing with Beaver Zion Lutheran. To that extent, the Kohn farm laid claim to their loyalties in the visceral way that sociologist Ferdinand Tonnïes calls *Gemeinschaft*. Dale, the youngest, did a tour with the Air Force and was assigned to the U.S. space program south of Houston, where he elected to buy a house and start a family. He later was joined in Texas by Harvey, who moved his wife and children there because of an excellent offer from a Miller brewery. None of us became farmers, but if anyone broke completely with my father it was me. The organizations and places that laid claims on me—Students for a Democratic Society, the Washtenaw County divorce court, the San Francisco hippie scene, Hollywood, the Los Angeles rock world, *Rolling Stone* magazine, the New York publishing establishment, and so forth—was as foreign as could be to my father.

I was not only another runaway, but, as someone wantonly and hugely malcontented, I was the flipside of my father. Where he was rooted in the earth, I was on the loose and on the make. And each day I was away from the farm, the more incomprehensible my father's life seemed to me. Even in the indirect, Biblical sense, what was it?—really no more than a delaying action. Sooner or later some big, rich farmer would take over the Kohn land in synchrony with the larger twentieth-century appetite for acquisitions and economic cannibalism which, let the arbitragers teach us, is at once the spirit of the American dream and the essence of one generation's superiority over the last one. If my father had been at all predatory, grabbing land when neighbors died, he might have had a fighting chance, but, on his one hundred and twenty acres, he was in a state—of ennui? of hiding from the real world? of quailing before it?—of having a total disregard, at any rate, for bettering himself, so it seemed to me. He represented the great romanticized myth of the American farmer, trying to survive on his terms no matter what. I knew the myth, of course, and I had marveled at it, been staggered by it, but I had never believed in it.

My leaving was a psychological and cultural flight, virtually a denial of where I was from. Over the years, on those visits that I did make to the farm, there were things I wish my father and I could have said to each other about amnesty and reconciliation. But we did not talk of such things, and perhaps it was just as well. Years earlier, while I was still a teenager, my father threw me out of the house ("I don't want you back here until you can act like a member of the family"), and when I did come back, after a period of rooming with high school friends, neither of us said a word. Silence was the best cover for our feelings. Then, that first evening at the hospital, in my fear over his condition, I nearly said something once or twice, something like "I'm sorry you didn't have better luck with me," but I did not quite manage it. A few days later, when Diana and I took our leave, after he had made it obvious that cracked vertebrae did not matter to him and that he was not going to be kept down, my feelings were different. I was angry that he could be so cavalier about injury and medical opinion, and yet—and this was the ironic crux of our relationship—I was grateful. It meant that he could go on with his life on the farm, and I could go off to mine, and we could act as if neither of us would ever have to change. A few months later my father wrote in a letter, maintaining the pretense, "I'm back in the old groove again. The doctors have had to admit that I've got a 'durable' body."

ONE

November 11, 1983

Hi! Howard and Diana

It's a rainy day, good to be inside. Have all the fall work done, finished the soybeans and corn last week. Don sold the corn Monday morning. The price was down from last week but we still made out pretty good, better than the last few years. That big apple tree on the front lawn broke over. Finished cutting it up yesterday. Almost have enough wood made for next year. About cutting up wood—how long will you be staying? Roy has a big pile of wood to be buzz-cut, and we could use some help with the logs. They are drilling an oil well ¼ mile north of our place. Guess Michigan is going to look as classy as Texas. If we don't hear from you, we'll be expecting you next week.

With love,
DAD AND MOTHER

....

The Saginaw Valley at night! Big changes can be missed by someone driving through. The oil rigs were new. The radiation detectors atop electrical poles—an early-warning system for the Midland nuclear plant—were new. A monolithic brick addition to the global business of Dow Chemical Company, set well back from Carter Road in old cow pastures seven miles from my father's farm, was new. The rubbled pits from the demolition of old farmhouses were new. Diana and I drove past and saw nothing new. She was asleep next to me, and I was tired after the long drive, thirteen hours long, from our house just outside Washington, D.C., and was irritated over our late start that morning. I kept my eyes on a long, open stretch of Carter Road lined by ditches. It ran through Beaver Township, the part of the valley that always looked the same to me. I was thinking of the time, several years ago, when Diana and I had turned off U.S. 23, then right onto Carter Road, and there, as we jolted along, dust spinning about, she had first seen the Kohn farm, in that same blank countryside, so lonesome, so empty of people, such a repetition of fields and woodlots and creeks filling with brush, so anachronistic, as if on Sunday mornings men and women in long coats still rode to Beaver Zion Lutheran behind draft horses. The farm was out where we had always said it was: beyond the back of beyond. At the last minute I had had sudden doubts about bringing Diana here. To a woman of schooling and big-city life, the farm would seem a dead end, with nowhere to go, nothing to do, no one to understand her, as it had felt to me. I might have hit the brakes and turned tail, but Diana had touched my hand. "It looks so peaceful," she had said genuinely, "exactly the way I imagined it." And over the years, our visits to the farm had been peaceful. Diana had enjoyed her talks with my mother and had found soft spots in my father. Yet every time I repeated the journey to the Saginaw Valley and made the turns toward the farm and became again more purely my father's son, I felt doubts. I was the exile who had scorned his homeplace, and all my homecomings were like the homecomings of poetry.

See, they return; ah, see the tentative
Movements, and the slow feet,
The trouble in the pace and the uncertain
Wavering!

On every visit I was older, but once I arrived I did not feel older. I felt like a boy. And I had the sensation of being in a single instant lost, undone, stripped of my accomplishments in the world outside the valley. It was a strange kind of traveler's anxiety, the only kind I had experienced.

Driving north now we neared the intersection at Beaver Road. I could always tell the intersection from the sound of it. Here the pavement of Carter Road gave out; the narrow ditched road became rocky; its bumps could not be avoided. With Diana's head warm against my shoulder, we crossed Beaver Road. I waited for the broken road to wake her. The only sound was the slickness of pavement. I found it hard to believe: the Bay County Road Commission's resurfacing program had reached the last, northernmost miles of the valley.

A mile later I slowed and turned off Carter. Then we heard a rough sound, the sound of gravel and dirt on the old slope of the farm driveway, eroded into parallel ruts, and we saw the familiar buildings, the red-brick two-story house, the mammoth gray barn, the sheds, the corn cribs, the chicken coop, all waiting across a long run of lawn. A few of the many fruit trees and giant shade trees planted in rows around the house still stood protectively, those not rotted and broken over, and it was possible to imagine, with some of their symmetry and grandness still apparent, how the farm could be a wonderful place to relax.

My father opened the storm door at the side entryway. "We went ahead and ate without you," he said.

"No sense everybody starving," I said back. My hand moved mannishly toward his, and he got a kiss on the cheek from Diana. He had come briskly down the steps, ignoring the metal rails that rocked at the push of a hand. He was more agile than most young men, and there was something ironic in his agility and in the energy with which he handled our suitcases, lifting them despite his back, almost in a parody, as if to ridicule the infirmities settling in, and perhaps to distract us from his own worry.

"How was the driving?" he asked, fishing for why we were late.

"Not bad," I said.

"As usual, we didn't leave Washington on time. It was after seven," Diana said in a straightforward voice that harbored an apology.

My mother held out her arms for us. But, catching sight of my father with the heaviest of the luggage—we tried to help, but he insisted on carrying the big pieces—her cheer faded. "Don't try to carry everything all at once," she said. Then to us, smiling gamely, "How was the road? How was the driving?"

In the kitchen the table was set, and the leftovers from supper were warming on the stove. The stove was a Kalamazoo, enameled white and shiny, with a big red bull's-eye temperature gauge on the oven door—a relic from the days of transition from wood to cheaper fuels, before the cycle completed itself and wood became efficient again. There were eight burners, four over spigots of gas, the other four, black thick metal lids, over wood in a firebox. A tank of fuel gas lasted my mother a long time. Even in deep summer her meals were almost always freshly done on old wood. The Kalamazoo, replacing my grandmother's black cast-iron stove, was one of the first significant improvements my father made in the house. He bought it the year Ronald was born and presented it to my mother. Nearly all the conveniences in the house arrived with her children, as part of the celebration. The year I was born (1947) a pantry wall was knocked down and the kitchen remodeled; for Ronald (1949), there was the Kalamazoo; for Harvey (1950), a refrigerator; for Sandra (1951), indoor plumbing—this because Sandra's birth was a hard one in a harsh month and the doctor ordered my mother not to use the outhouse; for Roy (1954), there was linoleum on the kitchen floor; and for Dale (1957), the Formica dinette set.

Diana and I pulled up chairs to the Formica table. My mother put out goulash and potatoes. They needed salt. "The shaker's in the cupboard," my mother reminded me.

"I keep forgetting." Salt was on a list of restrictions for my father, with his bad heart.

"Find it?"

"Right where I left it last year." I put a heavy dose on my potatoes, tasted, added more, a habit from my upbringing that Diana was trying to break.

We filled ourselves with homemade cookies and pudding for dessert, and, while my father warmed his back against the Kalamazoo, my mother rinsed our plates in the old porcelain sink with the scrub board on the side. Then all four of us sat on chairs near the stove. I noticed the time, past nine, the normal bedtime for my parents, but on the first evening of our visits home that deadline never counted. All the recent news was saved so my father could dispense it.

A few months before, oil had been found three thousand feet below the topsoil of Beaver Township, and that had brought in a rush of speculators. They had been knocking on doors up and down the country roads, asking to purchase an option to drill. "Now I've got two guys coming around who want to make me a millionaire," my father said. "One guy from Louisiana, he's disgusted with me. I think he went back to Louisiana. The other one's with a Michigan company. He's been here five, six times, and raised his offer to eight thousand, plus twelve and a half percent per barrel."

"Where in Michigan is he from?"

"Over by Mount Pleasant. He owns his own company, but he's a lease hound like the rest of them. Soon as he gets our section bought up he'll sell it to Sun Oil or some other bigger company. Sun Oil looks like they'll end up with most of Beaver Township. They've opened a new office in Midland."

"The ante will only get higher the longer you hold out. You're sitting pretty."

"I'm sitting right here—I'm not going to sell. What do I want to be a millionaire for?"

"So Diana and I can retire," I said, trying for humor.

My father's face permitted a small smile, which advanced to a chuckle. The chuckle was for his own joke to come. "The guy from Louisiana asked me why I won't sell my oil rights. He kept wanting to know why. I said, 'I'm saving them. I might need them some day.' And he said, 'Well, I've got to tell my boss something more than that.' I was going to say, 'Tell him it's none of his damn business,' but I said, 'Tell him, if he's so sure there's oil under my farm, then I'm going into the drilling business for myself so I can get a hundred per cent on every barrel.' He finally gave up. He said, 'Boy, you're one of a kind.' " My father's face opened again in a grin.

My mother looked at my father too adroitly. "He seemed like a nice young man."

"I don't mind them coming around, that's not it." My father rose, his profile telling sharply against the pale wood paneling, the full brow, the Roman nose, the stubborn jaw that always had tension in the set of it. With a portable handle he lifted one of the black stove lids and took the fire's measure. It had reconciled itself into ashes. He kicked them up with a poker. "But I don't like anybody who won't believe you if you say no." He intensified his labors and added a chunk of maple. "These guys, they think you're just trying to jack up the price. Their philosophy is 'Every man has his price.' I told them this isn't Washington."

I moved my chair closer to the stove. "Tired?" My mother meant I looked tired.

"Not yet," I said, but I was. The long drive had put me in a mood to go peaceably to bed. My father, on the other hand, would drive in sixteen-hour marathon shifts to Texas, to visit Harvey and Dale, and arrive ready to fix whatever needed fixing, savoring, with the proof in a long list of fix-it projects, the idea that old age did not necessarily require sleep.

One might not have thought right off to call him old. For the longest time, well into his fifties, his face looked remarkably boyish, and the secret to it was his apple-smooth skin that took a tan uniformly and his full blessing of hair that had held its dark color. Pushing seventy now, old age still seemed not to have caught him. His hair was graying but full. His face had the high color of a man of the outdoors. His eyes were like his complexion—a clear brown gaze. Fractionally short of six feet, he had a trim athletic build, strong in the shoulders and arms, with only a slight stoop from the fall off the roof. On evenings that he was dead tired, he may have felt mocked by his youthful looks, but tonight he was pepped up. He tipped his chair back and forth. He jumped up to attend to the fire. There was no end of talk in him, the kind of talk that we had precious little of when I was growing up. Instead, on free evenings like this, I kept my nose buried in *The Saturday Evening Post,* copies of which were delivered to us a boxload at a time secondhand from Aunt Linda and Uncle Henry in Bay City. Those magazines were the gift of fantasy. In the pictures and stories—we had no television—I saw far-off places that excited my dream of living elsewhere. (I have lived elsewhere since then, of course, and those places no longer seem so fantastic. Dreams are harder to come by, too.) After I left, I imagined my father's voice, with deep scriptural spaces between thoughts, reaching out to my sister Sandra and to my brothers, and I was glad, on the times I visited, when my father began to reserve the first evening for talking. The talks grew from the blurting of small gossip to an exchange of real happenings, grew into a tradition, and grew in my heart. So I knew to listen and forget about sleep. He talked about Beaver Zion Lutheran. In January he was to retire as chairman of the board of elders. "Just in time. The next year you can bet there's going to be a lot of arguing over the centennial," he said. The congregation, incorporated in 1887, was on the verge of a decision about an ambitious plan to top off its first one hundred years. "The young members want to spend a million dollars to remodel the church and school, and us old-timers aren't so sure. A million dollars! They better hope that every last one of these oil wells around here comes in!"

Moonlight stole in and picked up the glint in our unwashed silverware on the sink. My mother shuffled to the stove and examined the kettle boiling there. I half expected her to pour out a pan of dishwater. Instead she sat back down and interrupted my father. There is a low tone that is for her the tone of regret, and she said, in that tone, "Don won't be back with us next year."

"Why not?" I did not understand.

''Because—'' My father startled me by clearing his throat as my grandfather used to do. ''Don had his chance for five years. I told him that was long enough.''

Unsure of what to say, I said nothing. For a few seconds there was absolute silence. Don was Don Rueger, a cousin on my mother's side, born the year after me. An only son, he had been a junior partner on his father's farm, and now that his father, Harold, was nearing retirement, Don was the senior partner. During twenty years Don had enlarged and modernized the Rueger farm operation. It was the envy of anyone who had attempted to do likewise. At age twenty-two, Don had been the Michigan Young Farmer of the Year. For every phase of the farm season, plowing, disking, harrowing, planting, spraying, cultivating, harvesting, he had purchased mechanically superior equipment, the basis of championship agriculture. Each year his acreage expanded toward a goal of one thousand. As older farmers gave up and younger ones went broke, Don had more parcels of land available to him. With certain farmers he had handshakes that laid out the future. Ed Reichard, a neighbor of my father's without anyone to inherit his eighty-acre plot, was cropsharing with Don, and upon Ed's death Don would have first opportunity to buy the land at the prevailing rate. Five years ago, under somewhat comparable circumstances, my father had put his front eighty into cropsharing with Don. This had been a double concession for my father: that the full schedule of his farm was beyond handling by himself, and that it was a foregone conclusion none of the six Kohn children would step in and assume the load. He had accepted Don's terms, the conventional terms of cropsharing, three bushels to the worker, one to the owner, but the handshake between them was good only one year at a time. From the beginning Don tried to talk him into a longer commitment, ideally a signed contract: (''That's like asking for the moon from most older guys, not just him. They don't like to see you making plans for their land. Of course, that's what I need to do. I need to know I can plant *x* number of acres for *x* number of years, so I can figure out when to buy new equipment and how to amortize it. And how to plan my circuit.") Because Don did not farm contiguous land, but eighty acres here and a hundred acres there, he had worked out a circuit, like a picture of connected dots, that took his farm machines efficiently and progressively from one place to the next. I had watched Don and my father plowing one afternoon on the Kohn farm. They were on separate tractors and on opposite sides of the creek that divides the farm into a front eighty and a back forty. Pulling a five-bottom plow, Don was seated inside the air-conditioned cab of a late-model, eight-wheeled John Deere tractor, trailing diesel smoke. Superweight machines, unaffordable unless used on a large scale, were a strange sight on the Kohn farm. My father was pulling a two-bottom plow with a Massey-Ferguson 220. The tractor had four regulation tires and a pipe, attached to a rear-wheel hub by U-bolts, for inserting a beach-variety umbrella to keep bad weather off his head. It was purchased in 1966, before tractors were glamorous and when fresh air was thought unavoidable in farming, and was of compact size, appropriate to my father's status in the situation. My father was running it slower than the John Deere. Don had a schedule to meet; he would stay past midnight, headlights blazing, until the front eighty was entirely turned over. The two tractors moved back and forth—folklore confronting progress—the prelude, you might have guessed, to an inevitable falling-out between the two drivers.

"I was going to wait until the new year, see how I felt then, but when he came over last week, I figured I might as well tell him right now I was through cropsharing with him," my father said.

"What did Don say?" I asked.

"He got a little upset. He was over here to combine, and he said he wanted to plant corn again, and I told him no."

"And he got upset?"

"He wanted to know why. I told him, 'Don, there's no use talking about it.' "

"He hasn't talked to us since," my mother said.

"Yes, but he finished the combining." My father turned to

look at my mother. ''After the new year I'll go and talk to him.''

''Maybe you can strike a new deal with him for next year?'' I said.

''No. I should've never started with him in the first place. That's no way to farm. I'd be better off doing it myself.'' I could see that the idea of having his farm under his control again had given my father new fire, at least for the moment. It might be said that it was a Germanic moment because control is what matters to old Germans. Control can bring out the best in them. They make use of it to demand incredible efforts from themselves. Last fall, on the back forty, after my father had mechanically picked all the corn he could, when it appeared he would have to leave some to the crows—about two acres that were too soft from rains to hold his tractor—he went in with a bushel basket and twisted off the cobs by hand. Each step he took was in clinging muck. There were more than a hundred bushels. He had to carry each one several hundred feet to a wagon on high ground. ''No one was going to do it except me and Ma'' —my mother had helped—''so we did it ourselves,'' he said to me in a phone call. He has often said this: ''We did it ourselves''—and always in the most matter-of-fact tone. It had taken a week, and afterward he had to handle the cobs several more times. They were spread on the barn floor to dry, forked into baskets, run through a hand-turned sheller. Then, oblivious to his investment of work, he did not sell the corn. He had his pride. The corn was a little moldy. He ground it into chicken feed.

On a small, self-sufficient farm, personal standards can be very high. In modern farming, the type of farming my brothers and I had wanted for the Kohn farm, the type Don practiced, the practicalities are different. Don had no time for picking, husking, and shelling corncobs individually. His waterlogged acres either would wait until a winter freeze, when the corn might be harvested and marketed regardless of mold, or the corn would go to waste.

Poking again at the fire, my father said, ''I don't want hard feelings with Don. I'll talk to him. But I'm not going into another deal with him. I'll find some other way. Don't you worry about it.'' He gave me a reassuring smile, although to me it looked like it was set in lines of doubt.

He shifted his weight and looked toward the clock. ''Holy cow, it's midnight. We better get to bed.''

''Yes, everybody's tired,'' my mother said.

I went upstairs and pulled a heavy quilt over me. The news they had given me to sleep on was more likely to keep me awake, though. Diana made a soft noise climbing in. ''What's going to happen now?'' she wondered out loud.

....

Nothing appeases the scared kid in us more than the rising sun. It rose over flat pillows in the southeastern sky. My mother, alone in the kitchen, was deboning a chicken she had cooked before dawn. Everything but the bones went into a beat-up enamel pot for chicken noodle soup. My father was busy outside with the chicken chores. ''We didn't save you any pancakes,'' she said. ''After all these years I've finally got it straight that you don't eat breakfast anymore.''

''I think I gave up breakfast when I gave up milking cows.'' I laughed lightly, but it was true.

My mother cleared a space on the table for two dishpans, which worked like a twin-basin sink, one for washing, one for rinsing. The breakfast dishes and silverware slithered in, and she poured boiling water from a kettle over them. She scrubbed and set four plates, a finger between each, in the rinse pan. Again the kettle, aimed with a deadeye: the water missed her fingers and hit the plates with a steady, relaxed force. ''There's something you ought to know,'' she said. We heard my father come in down below at the basement level and stamp snow from his boots. ''He has to go to the doctor every week for a checkup. All through this fall his blood pressure's been shooting up. The doctor prescribed pills, but it keeps going up anyway. It's never been so bad. And at nights sometimes he gets aches at the back of his head, so bad he can't sleep. He's killing himself, Howard. All this work, all this worry.'' More stamping from my father. ''Don't say anything to him now. You know how he is.''

I nodded.

A gray head of curls, done up in a home permanent, and a color-splashed blouse gave my mother an appearance of frivolity totally at odds with her words.

In one of her scrapbooks there is a Brownie snapshot of her, taken about the time she met my father. She was twenty-two, he twenty-four. It is possible to see an ultrafeminine figure, seeing her, as it were, through his eyes. She is posed with a motorless lawn mower, her hands on the push handle, but she is not pushing. She is outfitted in a long skirt of the 1940s that begins high at the waist and expands over the hips. It is proportioned for her, flattering her figure. She has paused from her lawn cutting to stand for the picture, and the next minute she will be back at work, but from her pose, the shapely skirt, a modest little tilt to her head, she might be going to a dance. Her eyes, expressive, cleanly lit, look out fondly, as they did now at me. But now that face with the periwinkle blue eyes was the solicitous, enduring face of the medieval woman in the wheat field with all the woes of the world lashed to her broad back.

My father's footsteps came up the back stairs and onto the back porch that he had enclosed with red siding and a row of picture windows. My mother lowered her voice and said something that was lost in the flourish of my father's entrance. He had an armload of firewood. ''Oh, good. I want to get this house heated up today, with Howard and Diana here,'' my mother said enthusiastically.

''Good morning,'' my father said to me. His face was red from the weather. He looked at the clock—almost eight. ''Must've slept pretty good in that old bed.'' He pulled out his handkerchief in a ball, shook it and wiped moisture from his glasses. ''Chickens aren't laying,'' he said to my mother. ''Only got eleven eggs.''

''It's the cold,'' she said.

"Guess I can check on them later." He took a few steps on a path of old newspapers put down over the linoleum and found a kitchen chair with a straight, hard back. The easy chairs in the living room were too easy; once in, he could hardly get out. The bottoms of his boots were wet, and the wet spread in an oblong on the newspapers. The old headlines grew vivid again.

Scrubbing and rinsing, my mother bobbed out of the way of a steamy cloud and acted like she was having fun. A strange masquerade had been imposed on her.

In nearly forty years of marriage, she could count the occasions on which she had been apart from my father—the six times she was in Bay City General Hospital to give birth, the time he was in the hospital for his back, the three trips he had taken as a lay delegate to Lutheran synodical conventions—and that was all. They were married in Trinity Monitor Lutheran Church, her home church, on New Year's Day, 1946. They had begun courting five years earlier, and, I thought to myself, I would have liked to have known them in that period, on the evening when my mother, who was then Clara Buchhage, a farm girl living in Bay City, hitched a ride from a girlfriend in a running-board Ford. Clara sat in the backseat of the car, the girlfriend in front with her date. He was driving; it was his car. "Hi." He reached a hand into the backseat. "I'm Fred Kohn." The front-seat romance did not last. Fred's girlfriend was informed by a fortune-teller that her husband-to-be would ask for her hand in six months, and Fred made no move within his allotted time. ("I was still sowing wild oats. And there was the war with Germany. I figured we'd get in, and I figured I'd go.") Fred took an interest in Clara and invited her to church socials, the Wenonah amusement park, and the ice-cream parlor, but on other evenings he took other young women, and so it went, off and on, for some time, until Fred, running out of excuses and girlfriends, finally came to see that Clara was the best match for him.

....

My father, restless in his chair, was soon back outdoors, loading his pickup for tree-cutting on the back forty. I went with him. The sky was big and full of nothing. Winter seemed to be present not so much in the snowy fields or the absence of bird songs as in the blankness of the sky, which told of a year almost expired. Winter flattened the view. The farm was blurred by the snow that lay over it, something huge and clean. The barn and the sheds stood out like monuments, rectangular buildings with pyramidal roofs, built for the generations. Their rafters and crossbeams, the wood in their walls, their wooden doors, their window frames, gray now, petrifying, had come from trees that were lumbered out on the arrival of Heinrich Kohn, a German immigrant of the late 1800s. Heinrich and his neighbors had brought down the top rank of hemlock and oak with two-man crosscuts. For planks and finer boards and studs, they hauled the logs to a sawmill; the rafters and crossbeams they fashioned themselves. They used broadaxes. They debarked, trimmed, squared, notched, and then lifted by pike poles the massive crossbeams, fifty and sixty feet long, upward into the structural barn. The timbers had stayed in place. The two-man saws were in a toolshed; my father and mother used them on occasion in the woodlot Heinrich had preserved. After a hundred years of cutting down trees on the farm, four acres of woods were left on the back forty. To get there we could no longer drive down the cow lane through the eighty and take the rounded concrete bridge across the creek. The lane was plowed under in 1978, shortly after my father sold his herd of sixteen dairy Holsteins, when Don was asked to crop-share. Until then the eighty had been fenced into a grid of fields and pastures, and shade trees stood along fence lines for the relief of the cows. But the fences had to be taken down and those trees uprooted. Don's machines operated best in wide, open spaces.

My father and I drove south on Carter Road to the corner, left on Seidlers Road, east three quarters of a mile, across the creek by the road bridge, and left over a culvert onto the forty, which is the top portion of the ┏-shaped farm. We drove along the creek. In the distance was a poplar tree, sagging into the creek, fifteen degrees from vertical, abrupt, standing alone, every branch visible. This single tree gave cinematic frame to the fields and took away the emptiness. A giant, a landmark—it was suddenly hard to see anything else; it grew, filled the windshield. It was not neat and austere anymore, but a tangle. My father veered the truck away, toward four acres of woods in the farthest corner of the forty.

At the edge of the woods he parked and shut off the motor. Over everything came the winter silence, an immense, prevailing, wraparound silence. Even the cawing of crows overhead made little impression; they could have been miles above us. They flew off toward a cornfield. Crows have a good memory and sharp eyes, and, without circling, will fly directly to a spill of corn on the ground and begin foraging. When a person in this part of the country wants to express a high opinion of someone's veracity he may say, "He talks straight as a crow flies." This is said about my father.

Off the pickup bed he took a sixteen-inch chainsaw, two axes, a can of gas-oil mix, a can of lubricant oil, and some rags. He divided the load, leaving for me the lighter stuff.

"How's your back?" I asked.

"It gets stiff. I can't sit any one place too long. Which is good. Keeps me moving."

He walked fast, chainsaw swinging from one hand. In his determination, he snapped dead branches underfoot, scattering the silence. Brush whipped against our pantlegs. The farm and my father have always resembled each other in their orderliness, but the brush represented disorder. Without cows to keep it trampled and chewed down, it was traveling, thickening, beginning a new forest. "Let's clear a patch out. We've got a few extra minutes," he said, and set down the saw. A blunt-edged grub hoe is the proper tool for brush, and, returning to the truck, he brought one out and handed it to me. I swung the grub hoe and hit frozen grit. Using an ax, one hand partway up the handle, he was hitting only wood. The other hand was wrapped around the base of each sapling, steadying it. The ax blade struck inches from that hand. He bent over, straightened up, bent over again. ''Okay, let's take a breather,'' he said. He leaned his ax against a tree and threw the cut brush in a pile on top of a year-old stump.

We were doing more than making the woods look suburban. Later in winter, with the brush as kindling, my father would fire the stump, and in spring he would rip out whatever did not burn. Near the truck there was a strip of cleared land, about four acres, that fifteen years ago had been woods. It had taken my father, with chainsaw, ax, matchsticks, plow, and his own sense of pace, those fifteen years to remove the natural vegetation. Left alone for another fifteen years, the strip would again be strapping brush and saplings; in thirty years, it would be hardwoods.

Had he wished simply to improve the market value of his farm—cutover land sells higher—he could have rented a bulldozer for a week. He had in mind a different plan, at once simpler and more roundabout and more in line with his concept of managing the trees, a management similar to the staggered harvests in the vastness of the U.S. Forest Service, but adapted here to a homelier woods. His woods was firewood, and firewood was the principal heat in the house. The four acres of woods left to clear were fifteen years of heat. Germanic farmers and farm wives are known for long lives, and my father and mother might well live another fifteen years. He had taken all this into account. ''Long as we're alive, we shouldn't have to buy wood,'' he explained. ''I can cut up five or six trees a year and have plenty. And if I clear out around these bigger saplings and give them room to grow, the grandkids can come here for their firewood after I'm gone. Even some of these elms might make it.'' The American elm all but vanished in the 1960s with the accidental importing of the Dutch elm disease. My father had brought down his great elms one by one, salvaging some of them for cheeseboxes, until every one was gone. Today the disease is permanent in American forests, but he had noticed that elm saplings were beginning to grow for five or ten years before the Dutch spores found them.

With his chainsaw, my father addressed a storm-mangled

white ash tree. He revved up the saw. A wedge fell from the tree, and the trunk had an old crone's smile. The smile looked east. Opposite it and above it, on a downward angle, my father described an arc in the trunk, pushing the chainsaw blade, tilting it, not letting it bind up. The white ash dropped east, falling past trees in its shadow out into open ground. There was some chance in the placement but not much. My father had cut the wedge just so, with his lumberjack's eye, and within a foot he knew where the tree would come down. He took an emery stone from a pocket and put an edge back on his ax, testing with a bare finger, and we began to trim limbs from the white ash with a slapping-at motion. The limbs made a kind of flat canopy over the ground. I trimmed branches from the limbs, which were salvageable firewood, and he chainsawed the trunk into Kalamazoo-length blocks. Then we sat. Two blocks became chairs.

"When is your next appointment with the doctor?" I asked.

"So you heard about that?"

"Well, you've got everyone worried."

He attempted a smile. "Tell everybody to stop. Worry is what kills people. That's what the doctor tells me: 'Stop worrying, and your blood pressure will go down overnight.' "

"Good advice!"

"Sure it is. Look at Reagan. You can see he doesn't lose much sleep."

It was his casualness, more than his blood pressure, that was disconcerting. The white ash block began to feel frozen. I had to stand. Beyond the chaos of the cut-up tree and our bootprints was a sandy rise, and at the crest, no more than a hundred yards from the woods, was the Beaver Zion Lutheran cemetery. It once had been part of the farm, and it was like the farm still, and was like the community, full of old things. Three generations of Kohns were buried in the cemetery, and spaces for my father and mother were marked out. What there was to see of the cemetery, from where I was standing, had plenty of room to be seen in, the sentinel trees all sparse, the graves all plain.

"Well," my father said, "you ready for noodle soup?"

....

Perhaps the happiest I had ever seen my father and mother was at the Beaver Zion Lutheran school auditorium the evening of their thirty-fifth anniversary supper party. The auditorium was filled with a throng of relatives and friends. Reverend Leroy Westphal was the emcee, and he made the party into a kind of roast, getting several of us to tell stories on my father. Needling on this scale had never happened to him. How would he take it? I sensed hesitation in the first person who rose to speak. Then, abruptly my father stood up, his face an offended mask that broke, just as abruptly, into a wild grin, setting the tone. He began to tell stories on himself. He revealed himself to have been a rakish young man. He told how he had first made eyes at my mother while on a date with another young woman and how, when he enlisted in the Army after Pearl Harbor, several young women promised to write and wait. "In boot camp, before we were shipped out, we had two or three leaves, and when guys came back, just about everybody had eloped or at least had gotten engaged. When I came back, they'd ask me, 'Well, did you pop the question yet?' And I'd tell them, 'Nope, I'm still playing the field.' "

Reverend Westphal, when it was his turn, gleefully told about one of his first encounters with my father. "To tell the truth, it wasn't exactly *with* him. I'd just moved here, and, as you know, Fred's place is right over the hill from the parsonage. I looked over there and saw he had cows that were fat and happy, very well fed. The next time I saw those cows I had an even better look. They were in my garden, eating my sweet corn. I thought, 'No wonder they look so well fed.' "

There is considerable admiration and easiness between Reverend Westphal and my father. Later in the evening Reverend Westphal congratulated my father for holding a series of high offices at Zion and said that a lot of church members were grateful. ''And a lot wished I'd stayed to home,'' my father shot back.

What stuck in my mind, though, was a comment in a serious vein that Reverend Westphal made as we mingled about the auditorium. ''This is the best medicine in the world for your dad, having his family all together,'' he had said reflectively. ''He's laughed more tonight than in the whole time I've known him. He's not one to complain, but I think he and your mom can use more excuses to get out and about. I don't have to tell you how lonely a farm can be.'' And this point came sharply back to me on the evening after our day in the woods, when Diana and I visited my sister Sandra.

Sandra and her husband, Mike, had had a brief flirtation with a seaport life years ago, but since 1971 they had lived with their two sons, David and Scott, in a blue frame house on Seidlers Road, a mile and a half from the farm. ''Daddy and Mother being alone, it's not a good situation,'' Sandra said. ''Not the way he is right now with those dizzy spells and that throbbing at the back of his head. Anything could happen.''

''He looked really good last night. And today, too,'' I said. ''Practically on top of the world.''

''He hides it. Especially from you, I think. And he has his good days. He was excited yesterday, I know. He's always on top of the world on the days you come back.''

''I see,'' I said. How life turns!

''How long will you be here?'' Sandra asked.

''Not long, less than a week. I have two articles to finish by December fifteenth. But I'll be back before Christmas. We can get together with Ronald and Roy and everybody and talk then. Maybe figure out a new plan for the farm, since it looks like things are definitely over with Don.''

''We can talk, sure, but there's not much we can do, realistically.''

''Well, he can't farm the whole one-twenty by himself, not as long as he insists on hoeing every field two or three times every summer. The forty is too much, for that matter. Somebody has to pitch in. Maybe we can take turns," I said.

"We're already doing that. Ronald and Roy take turns with the heavy work. They both helped dig the well deeper this past year."

"Oh, yeah. The folks wrote me about that." Digging the earthen well is dangerous work: release the force of an underground stream and the earth may shift, the walls of the well may collapse. There had been some anxious moments up above.

"And that's just one thing," Sandra went on. "There's always something. The boys, David and Scott, go over to hoe. Daddy gets them out there, just like when we were kids." Sandra corrected a strand of hair. She had a great deal of our father in her, the handsomeness, the direct approach, the confidence and stamina—here she was, with two boys, a full-time job, a large garden, all manner of church and civic activities, and a husband who had returned to school on the GI bill and was gone five nights a week on a campus seventy miles away, yet her house never an embarrassment if a member of the Ladies' Aid dropped in. "I try to do my share on the farm, too. We all do, those of us around here."

"Yes, I know." I felt a sudden regret for something long after the fact. Living my traveler's life, I had thought of myself, how lucky I was, and not of the people left behind. Sandra had the courtesy not to throw it up to me, but once she had said, remembering my departure as a big event in family history, "I guess I was going on fourteen, and I thought to myself, 'Yeah, he's lucky. He got away. But now it's going to be twice as hard for the rest of us.' "

I had taken for granted that Sandra and Roy and Ronald would spend their free days at the farm, as I took for granted that Sandra had assumed my place in the family over the years. She was the first to be told of my father's heart attack. He had named her executrix of his will. She was the one he talked to, the one my mother talked to. Yet every time I appeared here, Sandra allowed me back into the role of eldest son. It made my Kohn identity seem whole and lasting, always there. It was an illusion, of course, like my artificial comings and goings.

....

The next morning, when a pin fell out of a gas spigot, releasing a spring, opening a valve, shooting flames from the Kalamazoo, my father found a cotter key and slipped it into the pin hole. The cotter key was wrapped in plastic. Several years ago a Chevrolet dealer had mailed it with a flyer: THIS IS THE FIRST PIECE OF YOUR NEW TWO-TON PICKUP. My father had saved the cotter key: someday it would come in handy.

My father took the opportunity to repeat a saying of his: "Don't ever say a thing is broken. Say it's worn-out. If it's broken, it's your fault. If it's worn-out, that means you've worked hard." He said this while going out the door to do his chores.

Minutes later, a car pulled off Carter Road and bumped into the driveway. A stranger came to the door. He smelled of cologne and had on a wide hat, riding low. With a touch of his hand, the hat tipped grandly toward my mother. "How have you been, Clara? Talk any sense into Fred yet?" The man's face expressed itself with a sunny, practiced smile. I knew him then: one of the oil-lease hounds.

My father approached from the chicken coop with a pail of eggs. "Howdy, Fred," the lease hound said. "What're you doing out in this wind? It'll blow you away."

"Too bad your company can't harness it. Lots of energy in wind," my father said.

"Oh, hey, they're doing that in California. Wind power—one of the energies of the future."

"Like they say, history repeats itself." There was an edge of sarcasm in my father's voice.

The lease hound was confused. "Oh, you mean windmills. Sure, you used to see two or three to a mile. But, face it, Fred, they were always undependable. The wind stopped blowing and you were out of luck. Soon as the electrical lines went through, the windmills came down. How many do you see anymore?''

My father harrumphed in his throat but chose not to speak. He came into the house, and the lease hound followed. ''All right, Fred, what did I quote you last week? Eighty-five an acre? Well, I've talked to my people. They want to get this squared away, so they've given me the green light to pay you the full limit. One hundred an acre!'' He pulled a calculator from a shirt pocket. ''That's twelve thousand dollars right now. If that isn't fair, I don't know what is.''

My father was about to put a word in, but the lease hound, thwarting him, gestured at my mother. ''What do you think, Clara? Is that fair? It's a no-lose proposition. Whether the company drills or not, you get the money for the rights. Or, if they drill and hit oil, the royalty is twelve and a half percent.'' His pitch became more rehearsed, phrases and punctuation snipped from a manual.

''What if you drill and don't hit?'' I asked, with a little challenge.

''Either way, you get the twelve thousand. You can't lose.'' He stared at me.

''What I'd like to know is what you'll do when you're through drilling. You don't guarantee you'll put the land back the way it was. Your contract doesn't cover that,'' my father said.

The lease hound fidgeted. ''It's not in the contract, no, but we'll do that. You have my word.''

''I'm not calling you a liar, but . . .'' my father said. A mile south on Carter Road, on the old Horschig farm, a drilling crew had scooped out a square hole, about forty yards by forty and a third as deep. The drillers had bored through bedrock, surfacing nothing but brine and sulfates, and had moved on to another site. They had left the hole with a hill of dirt beside it, in full view of the road. How likely was it that the drilling crew would haul back an earthmover and fill the hole in?

''I wasn't born yesterday,'' my father said. ''I've done a little investigating. Last month, over by Midland, one of the seismic teams went through with their dynamite tests and set off a dynamite charge in a lady's front yard, and when she tried to get them to stop, they pulled out their contract and said, 'See, lady, you signed this. We got the right to do whatever we want.' " The lease hound said, "Well, ah—" but my father talked over him. "It's their attitude, and it's how they act—I object to both. This summer, we had drillers over near Beaver Zion, by our church and school. They dug a huge sluice hole, half the size of a football field, and filled it with water from our creek. They stole the water. And they knew they were stealing. They came at night and sucked it out into trucks. You could hear their pumps going full blast. And their trucks broke up the blacktop by the school. It got to the point where Eugene Behmlander, who lives over there, had to call the sheriff's department."

"You can't condemn everybody because of a few bad apples."

"I'm not condemning anybody. That's up to the Lord. But I'm not signing anything either," my father said. He had not planned on such a long conversation. He stood in the kitchen with the egg pail in his hand, dressed in his winter clothes, five layers deep, a T-shirt, flannel shirt, sweater, synthetic-filled vest, and a light jean jacket, a gypsy's outfit. "You'll have to excuse me. I've got to finish my chores."

"Okay with you if I leave the contract here? You can look it over."

"You might as well take it with you. Twelve thousand or twenty thousand, I'm not having any drilling on my place."

....

If twelve thousand dollars is well below the average annual income for the American family in the 1980s, it was, for my father and mother, about three times more than their constant one-year nonfarm expenses, which were the following: a Consumers Power electrical bill (kept to a minimum by keeping the lights off), a Michigan Bell bill (occasionally listing a long-distance call), an Aid Association of Lutherans health insurance premium, a fuel gas bill, certain groceries unavailable on their farm (including beef, pork and milk, since the cows and pigs were sold off), my father's blood-pressure medicine from Auburn Pharmacy, and miscellaneous items from K Mart or Penney's or 3-D. My father and mother owed no mortgage. They had no pleasure car anymore. The pickup and the farm machines were theirs outright. They bought everything with cash. They had no credit cards and no credit line at a bank; they had no checkbook; their money was in savings accounts. Suppose my father's combine broke down, a new auger was needed, the wheat was rapidly passing prime, a storm was due in, and his bank was closed—events wholly plausible. Either he would have cash on hand or an IOU would satisfy a farm-implement dealer. Strangers have insulted him by not taking him at his word. One time, shopping for a station wagon, he settled with a salesman on a figure that he promised to pay in full with a cashier's check the next day. The salesman demanded a token down-payment, though, and so lost the sale. "When I tell people something, I mean it," my father will say. "I don't like to have to tell them twice, because I meant it the first time."

I expected the lease hound to be back nonetheless. Having plunged thousands of speculative dollars into my father's section, Section 21 on the township map, his company could not accept "no" for an answer. Oil options typically have to be obtained in a bloc before they can be resold to a drilling company. One section—640 acres, a square mile—is a bloc. It is sometimes possible, under a legal provision related to eminent domain, for drillers to bring their rigs into a section that has not been fully optioned, but in Section 21 my father's farm was a large fraction of the bloc and probably had to be under contract before drillers would proceed. Otherwise they would bypass the section.

To be fair to the lease hound, he was offering my father top dollar. The twelve thousand dollars could have covered their expenses for some time and freed them, if they were willing to be freed, from the necessity of farming, for which the payback was more and more uncertain. Don had given my father an itemized list of income and outgo for last year. After subtracting seed, fertilizer, chemicals and gasoline, my father's one-quarter split was $4,600, two thirds of which was immediately set aside for property taxes. "You can't blame Don that the bottom's fallen out of farm prices, though," my father had said. "He and I don't have any dispute over money. That's not the reason I ended it with him."

....

Later in the morning my father spread his quarterly federal tax forms on the kitchen table. He double-checked his arithmetic on the back of a Publishers Clearing House envelope, which he had confiscated from my mother.

Of late, she was playing the mail-order sweepstakes that the rural postwoman brought—"just the ones you sign your name and mail in"—but even that was out of character. My folks never played poker or bingo, none of the fun small-stake gambling games, because, I suppose, they had to beat the odds on a far more serious level against the weather and machinery breakdowns and market prices. So my mother was defensive about her fling at the sweepstakes. "Who knows? It doesn't hurt to try and we could win," she had said, expecting my father to disapprove, which he did. "Where do you think those winnings come from? The corporations don't put it up. It's like taxes. They collect millions from all the Pa and Ma Kettles and give away the spare change. Sure, once in a while somebody hits the jackpot, but everybody else is out of luck." He had suggested—ordered is perhaps too strong a word—that my mother shouldn't be party to such an unequal redistribution of wealth. This morning, when he took the envelope, she had shrugged. Perhaps he was right. So far all she had won was a ring set with a glass stone.

At the table, she now was clipping coupons from yesterday's newspaper. My father wrote out more numbers on the envelope. He had a calculator in his desk, a gift from Ronald, unused, its batteries dead. "I need a glass of water," he said, and he brushed aside Diana's offer to get it for him. He pushed himself up from the table, so slowly he seemed to be growing. ''That's the hardest thing I do: sitting,'' he said.

''I thought it was doing your taxes,'' I said.

''That, too.''

In American history, it is likely to be a populist moment when you read of a farmer and his tax bill. Traveling on assignment in Nebraska earlier in the year, I had met Marty Strange, who had said, his eyes blasting into mine, ''Our tax system is an inexorable killing machine. Every year there are fewer family farmers alive to kick about it.'' Marty Strange, in abundant irony, grew up in Brooklyn, and in 1970 found himself as an emissary of the federal VISTA program in the shrunken town of Walthill, Nebraska, where today he is a director of the Center for Rural Affairs, dedicated to reducing or eliminating the federal influence on farmers. In Nebraska, and indeed in most of the Farm Belt, bulldozers were crawling about, smoothing the terrain to prepare it for thousands of center-pivot irrigation sprinklers. Level ground is required so the long, sectional tubes radiating out from a central motor can ride on rubber wheels in a slow circular sweep. Center-pivots are subsidized by the federal tax system and, according to Strange, have the effect of consolidating many small farm enterprises into a few mega-farms. The cost of one center-pivot system, which irrigates one hundred and sixty acres, is $60,000. Up to half is recovered through investment tax credits, and additional write-offs are gained by treating the cost of bulldozing as a capital improvement. After four years, land improved by one center-pivot increases in value by about $117,000—''net profit,'' Strange said. The Thunderbolt Ranch in central Nebraska, owned by Herdco Inc., a Denver corporation, had thirty center-pivots over five thousand acres. ''The same opportunity is beyond the means of a traditional farmer or rancher. He can't get that kind of start-up money,'' Strange told me. Thunderbolt is the name Herdco assigned to the land, purchased in 1979, that had been three separate, family-run cattle farms. Herdco's startup capital, I found out, was a $6.5 million loan from the Federal Land Bank, connected to and guaranteed by the U.S. Treasury. The case of the Thunderbolt, and, in general, the preference that the tax system gives to mega-farmers, brought into question the overriding aim of the federal government. Did the government consciously mean to hurry family farmers off their land?

The decline of the family farmer, of course, is taking place—and has been for years—within the context of a general decline of mom-and-pop merchants and community craftspeople, and any time good people lose out to large forces there must be villains. In the 1930s the villians in the Farm Belt were the banks and loan companies. In the 1980s they are the agencies and policies of government. My travels through the Farm Belt had been short, but long enough to observe a reemergence of populist fever, some of it violent. I had talked to a farmer who belonged to the underground group *Posse Comitas,* and who, apropos of nothing, perhaps to create a scene for me to write about, pulled out a shotgun and shot the insulators off a utility-pole crossbeam, showering us with sparks. Members of *Posse Comitas* were at the time implicated in five random murders. Several farmers had proclaimed publicly they were paying no further taxes pending a reform of the system.

In Beaver Township there was no organized protest movement. The gun racks in pickup trucks around here were only filled during hunting season. Oil strikes and the promise of instant wealth, fantasy taking over from logic, had done away with some of the anger that the hard times had caused, and it also is not in the Germanic character to find distant, anonymous villains. "The Germanic farmer is most likely to put the blame for any predicament on himself or on his family or neighbors because that is the scope of his world," Dr. Salamon has written. As for my father, he had a code that forswore violence and guns—only twice had he fired the family's twelve-gauge shotgun: once against marauding raccoons, the other time to kill a cow dog gone bloodthirsty—and that code forswore, as he said, "bellyaching." When I asked about the tax protestors, he replied, "They're helping the Russians and hurting themselves. The government will catch up to them and get its money. I'm not one who says abolish taxes. I'll pay my fair share, whatever the government says it is, though I'd be happier if I thought it was fair.''

Later he brought out slips of paper that showed, if not unfairness, then misapplied logic in his property taxes dating to 1952, the year the farm became his. ''Here's why everyone around here can't wait to sell off their oil rights,'' he said, giving me the tax bills to examine. In 1952 his property tax was $285, from which it increased an average of $19 a year for the next twenty-six years, until in 1978 it began to shoot out of sight on an average increase of $485 a year. The logic was that the good years of the late 1970s, when farmland values were spectacular, warranted a commensurate response in taxes—except now, in the middle 1980s, in the midst of the worst farm statistics since the Great Depression, land values cut in half, market prices fallen below farm costs, foreclosures and bankruptcies commonplace, the Saginaw Valley property taxes had not been adjusted downward. They continued to go up. My father's tax was $3,680. ''You don't make your loan payments, that's bad, but maybe you can get an extension. You don't pay your taxes, that's it: they'll sell your farm out from under you,'' he said.

....

Red and pink showed on the horizon. I came down the stairs into an empty first floor. I found my mother where I guessed she might be, in the garden, dumping out the metal grate of ashes from the Kalamazoo. My father was at his blood-pressure checkup. ''He wanted to drop off a few eggs at Maple's first,'' she said, ''and he has to be to the doctor by nine.''

''Has he been taking his pills?'' I asked.

''He takes them most of the time. But he thinks they might be giving him that ache in his head.''

I watched the snowy ground turn gray from the ashes. Thirty

or forty pounds, heavy with potash, would have been added to the garden soil by spring.

"How bad does he really feel?" I asked. "I can't tell."

"It comes and goes. The worst thing is he can't seem to sleep." She seemed to choke on the ashes swirling and giddying around her. "He doesn't tell me more than he has to. But I know he can't sleep."

My father would rise once or twice a night, she said, depending on the ache, which seemed to expand in cold weather and under stress. He woke one time shortly after midnight with a severe pain. "Will you get up with me, Ma?" he had asked. "I've got to sit till it passes." She helped him to a chair in the kitchen. She turned on the lights and brought the fire in the stove back to life and began to bake pies. He sat—did she imagine it, or was his color fading to lesser shades of gray? He refused to have an ambulance called, though. She baked twelve pies, eight apple and four cherry, and put them in the basement freezer. All night they stayed awake. In the morning he was all right.

"At least he's seeing a doctor," I said.

"At least he's able to do his work," she said. The obligation of work, regardless of his health, was a statement of the man and his culture. I remembered the winter that an extra job at the Monitor sugar-beet factory, taken to pay extra bills, had given him walking pneumonia. He spent his shift in front of a distillery furnace with a cold draft on his neck. Quitting was out of the question. Would he take a few days off? Out of the question—and, until spring came, my mother had worried herself into a scared-he'd-get-sicker, scared-he'd-cough-blood, scared-he'd-fall-into-the-furnace insomnia.

My mother and I were walking around the perimeter of the garden. Overwintering carrots were in the ground. A few black-purple Concord grapes hidden behind leaves at picking time stuck out now in the arbor. Grapes frozen on a vine sometimes grow sweeter and juicier, but these were bitter and cold.

....

For dessert that evening, the last before Diana and I were to leave, Sandra invited us over for cheesecake topped with hand-picked berries. Mike, home for the weekend from Ferris State University, showed off a large map of the campus, current and detailed to the last street sign. It was a class project. Mike, who had been a truck mechanic, was enrolled in a specialized high-tech program. When he was through he would know how to decode and draw maps from the computerized surveillance of high-altitude planes and satellites.

"Everyone who graduated from this program last year went right into a job. I'm hoping my class has the same luck," Mike said. He is soft-spoken—the antithesis of a braggart.

"You will, honey. I know you will." Sandra, on a brown-shaded Herculon couch, beamed over the map. She pulled me in for a closer look. "Some days it's hard to believe that this crazy idea of going back to school will be worth it, but when I see this map and realize Mike drew it, I know we did the right thing."

At a wedding nearly four years ago, in a lighthearted mood, with the family around, Sandra and Mike first toyed with the idea of changing over their lives. ("I was teasing Mike, 'You ought to go in tomorrow and quit and go back to school.' He'd had a drink or two, so he said, 'Okay, honey, it's a deal. You pay the bills and I'll go to school.' Well, the next day it still sounded like a good idea. Mike had eight years' seniority at Delta Ford. They thought he was crazy to quit; guys were waiting in line for his job. But it gets old as you get old, being a mechanic. You're craning your neck; he'd come home with a sore neck, a sore back. That's half the reason we bought the water bed.") Starting with a high school degree, Mike tackled two years of basic math and science at Saginaw Valley College while living at home, and now was a resident at Ferris State, an hour and a half away in pine country.

At Ferris State, the icebox leftovers that Sandra sent back with

Mike each week had not kept him in form. He had shrunk by fifteen pounds, although he persisted big in outline, six one, about two hundred pounds, with a tumble of brown hair, a walrus mustache. In 1970, one year after high school graduation, Sandra married Mike in the midst of plans meant to take them to naval ports somewhere warm and hospitable. Instead the U.S. Navy posted Mike to Bayonne, New Jersey. Sandra insisted on going along and found a clerk's job in a Bayonne shipyard. (''Foreign ships came into the harbor with words on them I couldn't read. It wasn't even our alphabet.'') Buses were her means of getting around. (''One morning I hopped on a bus marked EXPRESS, because I was in a hurry to get to work, and before I knew it I was on the Jersey Turnpike, going seventy miles an hour toward New York City. I said to the driver, 'Please, let me out. I'm on the wrong bus.' And he said, 'I can't stop, lady. You're going to see the Big Apple!' 'Lady?!' I was just eighteen. When the bus got to New York, he let me off. There were these guys on the street who—well, I was glad it was daylight. The next bus back to New Jersey, I was on it.'') Mike applied for and received an assignment to New Orleans, but somehow the transfer was misplaced in a computer, and the orders that ultimately came through were for Newport, Rhode Island. In Newport, Sandra and Mike were even more obviously rural Midwestern—overly inhibited, overly outgoing, out of rhythm with a sea-sporting crowd whose style was more Royal Navy. Soon Mike's duty ship, the *USS Paul,* embarked for the Pacific basin, and Sandra, pregnant, moved back into her girlhood bedroom at the farm. My father wrote to me making it clear that Sandra's adventure was over, that she was staying put until Mike disembarked and flew back to Beaver Township, that he was not reenlisting, and that there were no other destinations in mind.

Now, after distributing beer and potato chips, Sandra said she had an announcement. ''We need to buy a big map of the United States,'' she said, and, plopping onto Mike's lap, spoke more decisively. ''We want to move somewhere where the economy is strong, a place we can count on for a while.'' The 1980s had been tough years in Michigan, tough because so many in the state were dependent on either the automobile industry or agriculture.

"Have you told the folks you might move out of state?" I asked, then, sensing the question was tactless, added, "I don't want to say anything that'll put my foot in my mouth."

"No, not yet. I think Christmas will be a good time. Lots of distractions to keep them from being too upset."

I nodded, and we took another look at Mike's project. He overlaid the thin draftsman's lines with a computer printout of numbers, reading from a plane. The numbers dictated the elevations on the map. He held a pencil and traced a line to demonstrate. After years of fingers dirty with grease and thumbnails blackened by slippery wrenches, the abrupt stroke of his hand to the paper was straight and clean. It pointed toward the future.

....

Back at the farmhouse my mother was awake. She had decided to sew buttons on a knee-length leather coat that I had once picked up at a yard sale and left in an upstairs closet. "I think I have four brown buttons that'll fit," she said. She rummaged through a King Edward cigar box in which she had stashed buttons of all shapes: "Anything that's worn-out I always snip off the buttons before I cut it up for rags." One by one she plucked out four roundheads, miniature chestnuts, almost too fancy for the coat. "Do you like them? They match." I said I liked them.

Into the leather her long thick darning needle took a hard push. Even with a thimble her fingertips turned red. She sat at the kitchen table, concentrating on the needle and pretending for a few minutes, I felt certain, that I was a boy again. I had not worn this buttonless coat in years. What was the point of fixing it now? Diana and I would be back in a month for Christmas, but it was not a winter coat. It was a good coat for May planting, but I had not said a word about being here in spring, and my mother had never interfered in my adult life. She might simply have been saying with her needle, ''Your coat will be ready, just in case you happen to be here in the spring.''

The leather was resisting the needle. My mother pulled at it with her teeth, incautiously biting the point, fingers taut on the leather, and the needle came vehemently through. She tied off the thread. ''Here, try it on,'' she said. Her face was caught in an expression half prideful, half uncertain. I put it on, and she buttoned it. ''There, almost as good as new,'' she said.

TWO

Ten days before Christmas, in falling snow, Diana and I made the long drive back to the farm. On every previous visit the farm had seemed exactly as it was when I was young. The house and barn and sheds looked the same. With the consecration of time they had become fixtures, constant as breath and, what's more, taken as much for granted. My parents looked only marginally older. Even with the changes—the livestock sold, the fences torn down, Don's diesel tractor in evidence—the Kohn farm remained in a rural permanence under my father's firm hand. Life continued as before in the Michigan countryside that in summer, though Canada is not far away, recalls the rich green of the South. Every summer had its regimen, for my father more of a celebration, of walking the fields, hoe in hand, my mother alongside, barefooted in the warm dirt, her shoes left under a tree. Harvest followed harvest. The seasons repeated themselves. Nothing truly changed. Until now—now change was upon them. On our visit a month ago my mother had said, "The work is too much in summer. We can't do it all."

My departure years ago surely had prefigured this dilemma—but equally surely it had not. Tax inequities, federal subsidies of corporate farming, competition from foreign farms, the monopoly practices of grain companies, the fact that across America farming had become big business while on my father's farm it had not—all of this finally was forcing his hand. Had anyone inquired whether I felt guilt, the need to atone, I would have said the opposite was true—that if anything was working on me, it was not my past, but Heinrich Kohn's past, the imagination and the sense of possibilities that had led the Kohns to Beaver Township in the first place. Standing pat at this point could not save the farm; I knew that absolutely. No matter what, it was going to take an assumption of risk and some of Heinrich's original pioneer daring.

Before Heinrich ever boarded the cargo ship in Leipzig that brought him to America, he had in his head a map of his route from Germany to Beaver Township, a map associated not with ocean lanes and overland passes but with a vision. All he had reason to believe was that German Lutheran outposts existed in a place called the Saginaw Valley. His own prospects he had to guess at. And how would he find his way there once off the ship in New York? Who would guide him across the Alleghenies and the Ohio flatlands? Yet, in 1876, at age twenty-five, he packed up and sailed off to find somewhere, somehow, a farm. His pipe, his razor, leather boots, clothes, his Bible he brought along in a wooden crate on which his name was written in English. On board the same ship was Augusta Tomke, who was sixteen. Augusta had only those belongings that could be stuffed into a cloth valise. Her emigration had been planned in a hurry. She was traveling under another name. A German husband and wife—the woman, pregnant, ticketed, but already feeling nausea—had had to give up their passage and scalp their tickets. Augusta, a farmer's daughter, dowryless, her prospects in Germany unpromising, went in the woman's place, no doubt wide-eyed at her sudden ride to America. Augusta may have met Heinrich on the ship, although the family's oral history says she did not. They are said to have met for the first time in Bay City, on the West Side, in a neighborhood called Salzburg, a roughneck, itinerant, principally German area, where they fell in love. Two years later, years he worked in Bay City sawmills to build their savings, they were married. As I understand it, he left

Augusta at their boardinghouse room one morning and went exploring, venturing into the sparcely settled townships that lay west between Bay City and the next big lumber town, Midland. It was spring. The creeks were above their banks. On the muddy trails, wagon and horses bogged down. Heinrich returned to Bay City and rented a skiff that he and a friend poled up the Kawkawlin River, off which forked a fantastic, pummeling choice of byways, presenting him with dozens of unpredictable milieus for developing a new life. He poled as far as he could and then hiked to the northwestern corner of Bay County, two miles from Midland County. In the wildly growing green, he saw a place to make a farm.

When Diana and I had visited last month, a century later, we counted four pages in agate type of farms up FOR SALE or up FOR AUCTION in the *Midland Day News* classifieds. As things stood, the Kohn farm would soon appear on those pages.

No one asked me to come up with a plan of action, but in a notebook I had scribbled a few thoughts. To buy time, a year anyway, my father could enroll the farm in the federal payment-in-kind (PIK) program, which was subsidizing farmers to keep their fields fallow; alternatively, he could sell the farm's oil rights to the highest bidder, and, if necessary, he could sell off frontage. For the long term I had this idea: acquire a federal research grant, relying on friends I had in Washington, and turn the farm into a series of test plots, which my father could work in concert with a team from the U.S. Department of Agriculture's far-flung research system.

On this trip home, I thought, I will arrive at the farm finally grown up, ready with a plan—a vision—that proves maturity. Leaving Washington, I felt pleased. But the feeling did not last. I began to brood. After years of keeping my distance from the farm, what could I hope for now? Over the Alleghenies, down onto the Ohio Turnpike, north at Toledo, into Michigan on U.S. 23, I turned the question around in my head. If not atonement, what? The Beatles and the Rolling Stones reverberated through the mesh of the car speakers. "Why are you so quiet?" Diana asked. I tried to explain, but I could not be sure I was being honest. My plan included an ambiguity: Was I aiming to preserve the farm for my father, or did I want to assert my ideas for their own sake, to seize a final adolescent opportunity to brag and strut and prove something?

"My advice is that you don't go charging in like you've solved everything," Diana said.

Fields without trees and ditches overgrown with primordial forest surrounded us as we left U.S. 23, and surrounded also my father as he held open the door. Wind tried to suck it shut. The snowfall here was a storm. My father was in his pajamas. "Sorry to keep you up," I said. It was almost 11 P.M. We were five hours behind schedule because of the snow and delays getting away, a phone call from an editor, arrangements with a neighbor, the details of my life.

"Just bring in one suitcase. We'll get the others in the morning," he said. "Ma's already in bed."

....

The snowstorm went on through the night. I awoke intermittently. In front of the house the wind bent the branches of the trees, two maples and a white ash, and they scraped viciously on the slate shingles of the roof. The noise brought back the memories of all the other storms I had held my breath through, and the one storm that had snapped a heavy limb and driven it into the roof, scattering the shingles like playing cards. Above my head, the wind and branches were trying once more to break in.

In the morning, on the front lawn, I picked up the litter from the trees, busted-up branches: kindling for the Kalamazoo. A livid blue sky, almost hurtful, hung over the farm. The wind was gone, and the trees had a solemn stillness. In the sun the icy buds of the maples looked like steel points. A clothesline ran from one maple to a white pine that had been trimmed of branches and creosoted into a telephone pole. My mother stood next to it with a basket of clothes wet from her washing machine. Clothespins dangled on the line, heads pointed up, as if conducting an ancient form of worship. "Need any help?" I yelled. My mother shook her head. She pinned the clothes to the line. In an hour they would be freeze-dried. On the surface of the laundry, molecules of ice would jump sublimely into the air. Sublimation, or freeze-drying, by which a solid mass distills into the air without first passing through a liquid stage, is a phenomenon known to the Quechua Indians of the high Andes, who lay out potatoes to freeze and dry, and to Eskimos, who suspend salmon and seal meat on poles, and to the Lipton Company and the Oregon Freeze Dry Foods Company, whose dehydrated packages of soups and beef patties were carried by Marine patrols in Vietnam, and to my mother.

My father's voice boomed from the coop, past the corn crib and garage and toward the front lawn, and I continued to hear it until he saw I was coming his way. The snow made for slow walking. He was in a temper at the weather and at himself. "Damn chickens," he said. One was dead, killed not by the cold but by humidity and gases rising off the manure. Inside the coop, sublimation had come to a halt. A while ago my father had packed newspapers under the eaves and around the windows to raise the temperature above freezing. He had succeeded too well. A vapor was trapped inside, damp and penetrating and worse than the cold. I could see it leaking from a few cracks he had missed. It froze on the outer wallboards and at the edge of the door, forming complicated crusts of frost and bubbles of ice, one on top of the other like rhinestone jewelry: a pretty sight, a gas turning solid, the reverse of sublimation.

"We'll have to take this stuff back out," my father said. He yanked at the wads of newsprint. He looked unhappy and wadded up himself inside his layered costume, his breath smoking like a meat locker, his bright ideas over.

The chickens flapped their wings, scurrying, squawking "pa-ba-bawk," a noise the ancient Malaysians heard in their tropical woods, from whence came the modern poultry birds, by way of Europe, and, most recently, in the case of these chickens, by way of California and New Hampshire. They were a crossbreed of the California white leghorn and the New Hampshire red and were known as California browns. "More bother than they're worth," my father said. I had heard him on other days justify to my mother that the California browns paid for themselves, however. Each hen laid an egg on the average two days out of three, and he sold the eggs at sixty cents a dozen to customers in a free-lance network—to neighbors, relatives, several of Sandra's coworkers, and, for resale, to Maple's, a throwback grocery with two gas pumps on Garfield Road.

The chickens settled down, feathers whispering against feathers, their bellies pressing flat to the straw bedding. "None of them are feeling too good, and they're molting, which makes them feel worse," he said. "They're down to seven, eight eggs a day." In the good times of summer, their quota was seventy-five a day.

Wind was starting to blow under the eaves and into the coop. "Chances are, some will freeze to death," he said.

"How are you feeling? Any more of those pains or dizziness?" I asked.

"Me?! I feel fine!" The words were exclamatory. I suspected they weren't the full truth, but the fervor in his voice almost made them true.

At noon my mother hurried the soup, heating it on a gas burner, and it was scalding. We said grace—"Come, Lord Jesus, be our guest, and let thy gifts to us be blessed"—and blew the soup cool on our spoons. Sitting at the table, Diana's back was to a frosted-over window that let in air. The caulking no longer sealed the panes. "I hate this cold," she muttered. My mother looked less vulnerable to heat loss in her heavy sweater, but she was rubbing her arms through the yarn. She worried steadily that the cold would worsen. "Did someone see if we have enough wood for the stove?" She addressed all of us, and, no longer solicitous, said to me, "You should have brought some in when you were out."

"There's plenty," my father said. Finding the Kalamazoo in

faint embers after breakfast, he had restoked the fire and had resupplied the porch with four armloads of split cordwood.

The cold did get worse, and, except for a fast run to the coop before supper, we stayed in the house. My mother and Diana baked cookies—chocolate chip, molasses, sugar stars—and my father and I read in the living room, he his monthly copy of the *Michigan Dry Bean Digest,* I the newspaper. Our eyes grew tired in the weak light, and we put down our reading. Over many minutes we spoke once or twice. We watched the blank gray face of the television set and then switched it on to a ''M*A*S*H'' rerun. In 1963, after years without a TV, my father, with no warning, bought one, perhaps hoping that I, then sixteen and restless to the point of cruising and drinking until dawn, would spend more evenings at home, although, of course, I did not.

....

The next morning at six—what my father says is ''the middle of the morning, halfway to noon''—my mother lit a new fire. She began with corncobs and a sprinkle of kerosene. I was up early after another restless night. It was pitch-black out. I ate a single slice of toast that my mother buttered. For toast, before she had a toaster, we use to spear a long-handled, long-tined fork sideways into a slice of bread and hold it to the fire in the Kalamazoo.

''I hope things get straightened out with Don,'' my mother said. ''The way it is, I don't even want to call Thelma. I don't know what to say.''

Thelma Rueger, Don's mother, was my mother's first cousin, and they had been close friends since childhood. At my father and mother's thirty-fifth anniversary party I had taken several pictures of Thelma with my mother, chatting girlishly, their heads together or thrown back laughing. The two families were close, too. Thelma was my brother Ronald's ''sponsor,'' the Lutheran equivalent of godparent. My mother was sponsor to Don's sister, Joyce. The Kohns had spent many Sunday afternoons with the Ruegers. The afternoons clearest in my mind were in winter on a thin rim of crusty snow at the M20 overpass, near the Rueger farm, with Don and Ronald and myself on sleds, our younger brothers and sisters looking on and shouting for their turns. The overpass had banks that were practically vertical, and Don and Ronald and I drove our sleds straight down, marking trails that with constant friction dropped us faster and faster to the flat plain below—on a principle of ruinous acceleration that is easy to comprehend but was on such a breathtaking scale that we ignored it and had a ball until we were all slammed from our sleds.

"Is there anything I can do?" I asked my mother. "Should I talk to Don?"

"I wish I thought you could do some good. Maybe it's best to let it alone for now."

"You're probably right. Plenty of time before spring."

My father came in, holding a flashlight and his egg pail. He had been to the coop. "Six eggs," he said. "One's completely frozen. Solid as a baseball."

"I can use it for the meat loaf. It won't matter," my mother said.

"You'll have to peel it like hardboiled," he said. He smiled briefly.

"Any dead chickens?" I asked.

"A couple are halfway there. I might chop their heads off later today. Put them out of their misery."

In the afternoon my father and I carried burlap sacks of chicken feed from the barn to the old icehouse, where they would be more accessible to the chicken coop. We laid the bags, one at a time, over the hunch of our backs. The human back as carrier, the legs as transport—it is a practice of agriculture worked out over the centuries, learned so long ago and so well, one might say, that it cannot be taught today by words. It is learned by example and then by doing. The balance of movement is complex. Spread the weight assuredly across the shoulders and upper back. Crouch, lift with the legs, and lean forward. As in athletics, a rhythm must be struck. How many Americans can carry a feedsack on their back? Compare it to a modern hiker's rucksack with no aluminum frame or nylon straps. Imagine putting a file drawer of papers on your back, and the metaphor about "carrying your own weight" clicks.

My father and I kept at it, the heavy trudge through snow, a hundred yards each way. "Now you know why they say farming is a young man's game," he said with a slight wheeze.

I saw an opportunity to put forth part of my plan. "Have you ever thought of taking a year off?" I asked. "You could enroll the land in PIK, and let the government pay you not to farm."

"PIK is another one of those programs that's set up for the big corporation farmers, not for us little guys."

"I'm sure you'd qualify. Anybody who grew corn last year qualifies."

"Who would want to? If you're not going to work, you better retire."

"You could do that, too," I said, an unfortunate goading that slipped out.

He dusted chicken-feed powder off his coat. "One of these days," he said briefly. "Not quite yet."

The cold had the feel of imminent frostbite, and we had to keep moving. Conversation slowed us down and was a waste of energy, and, the more we talked, a waste of time. My father clearly was not interested in the PIK program. Or in selling his oil rights. Or in selling frontage. Two miles north on Carter Road a farmer had subdivided land along the road into long rectangular five-acre strips, hoping to attract Dow Chemical engineers who wanted to breathe country air. It was a largely failed endeavor my father referred to as "ridiculous." My long-term plan—to convert the farm into a kind of laboratory—I did not bring up. My mention of retirement put a damper on everything. My mother had told me that, after my father's heart attack, certain realities had hit home, but neither of them was able to picture him in formal retirement. "It's the work that makes him feel good," she had said. "What's he going to do if he doesn't have work?" Now my father made it a point to tell me about a young farmer, a "weekender" with twenty acres of oats and a few beef cows, who had noticed my father's Massey-Ferguson Model No. 72 eight-foot combine and stopped to inquire about it last fall. The combine, well preserved, was of efficient size for the young man's farm. He praised its condition, the smooth, greased turn of the flywheel gears. My father was bolting an Innes cylindrical bean pickup to the combine for harvesting navy beans the next day. He was courteous but not altogether friendly. He cut off the run of compliments, holding up his wrench. The young man became more businesslike. What might the combine be worth? "It's not for sale," my father said. "Give me your phone number, and I'll let you know when I retire. You can get it cheap then." While telling me this my father waved a hand, palm outward, as if to dismiss the scene.

....

On Sunday we attended the 10:30 A.M. service at Beaver Zion Lutheran. Snow had brushed the road overnight, and the sun made it look so hard and glittery it was difficult to see.

"Good morning, Fred. Morning, Clara," one of the elders on duty greeted them, and, indicating Diana and me, added, "I see part of the family is back for Christmas."

"They're all coming. Harvey and Dale arrive next week," my mother said. She had put on her gold band with the six birthstones, one for each of her children, special-ordered by Sandra a few years ago. It was the one gift I had seen my mother cry over.

The pantsuit she was wearing, practical, unfussy, would have caused a sensation not long ago. The dress code for women—no pantsuits, no slacks, no culottes—was an explicit part of church policy as late as the 1970s. I remembered that Dorothy Stieve, the Zion teacher who taught remedial classes, had to ask permission of the principal to wear slacks even to school. ("Yes, that wasn't so long ago. But we women wear what we want now. We can wear jeans to communion if we want. It was a long time in coming, like everything else here, because we're set in our ways.'') The first major break from custom came after World War II. Men who had seen London and Paris, and who had been separated for years from wives, girlfriends, and mothers, decided they were ready for the sexes to sit together on church pews. Until then, men and knickered boys had been seated to the left of the middle aisle, women, girls, and small children to the right. Women continued to have to wear hats in church through the 1960s. Nancy Bliese, my first-grade teacher, panicked some of the older men—and women—when she purposefully left her hat off one Sunday, about 1955, and, after a number of complaints to Reverend Fred Reimann, he announced in the weekly bulletin that he would skip over any woman who did not wear a hat while kneeling for communion. My father had said this about Reverend Reimann: ''He believed that if a woman recited the prayer before a meal the food wasn't blessed.'' Hats on women in church was not completely a dead issue at Zion. ''Periodically it surfaces,'' Reverend Westphal had mentioned to me. ''It's discussed because Saint Paul did say that for women to appear in a place of worship with their heads not covered is unbecoming.''

My mother, hatless, walked behind my father to a pew, and Diana and I followed them. My father had on a paisley tie and a striped yellow shirt and a brown sport jacket. In his wide, blunt hand, so controlled and natural upon an ax handle or hoe, he held his hat, unnaturally, as if it were tissue paper. It touched without weight the cloth of his brown pants. The hat was made out of a synthetic gray fur. A pheasant feather was tucked into the band.

The congregation rose for the liturgy. Reverend Westphal, in plain dark cassock and white surplice, chanted, ''Glory to God in the highest.''

''And on earth, peace, good will to men,'' the congregation replied in a cracked half-song that filled the big church. The words were perfectly familiar and came easily, inevitably, to me even though long ago I had stopped going to services.

Behind Reverend Westphal was a soft-white arched wall, backing a modernist cross of shiny wood and gold outline. There had been one reconstruction of the church, in the early 1960s, to please streamlined tastes: pews without ornate carving, an elegantly simple altar, clean wall hangings with blood red symbols on white cloth. As a young acolyte, I used to gravely approach and light the fourteen candles in the twin altar candelabra. For a brief time earlier in the century, excited at having the church wired, the congregation installed electric candelabras—"Goes to show that even the old-timers sometimes lost their heads," my father had said—but the traditional candles now were back in place. Two steps down from the sanctuary, the facing walls used to have murals, dimmed by candle smoke, of Jesus knocking at a door and Jesus as a shepherd, sandaled feet peeping beneath a long white robe. One Christmas, Sandra received paint-by-number copies of these Jesus illustrations, and, painstakingly filled in, they had hung in the farmhouse living room. Recently my mother had taken them down and given them to Sandra for her own home.

After the final hymn, "Faith of Our Fathers," which fell into a dispirited minor key here and there, and after the final Amen, we bowed our heads and Reverend Westphal walked down the middle aisle to the great oak doors in back. His face ruddy from the sermon, he waited to shake hands as we were ushered out. "Hello, Howard," he said to me, his grasp firm. "Glad you made it for Christmas."

Reverend Westphal, the thirteenth minister to serve at Zion, was a robust, intelligent man, the father of four grown or half-grown children. His shoulders and back and jaw were at right angles, his eyes level, looking straight at me. His hair was at salute length, but he had a grace and decency that saved him from resembling a drill sergeant. The son of a Lutheran minister, he was raised in an Illinois parsonage that had a farm out back. There were pigs, chickens, sheep, a horse, cows, and one of the cows was his to milk. There were no milking machines. The milking was by hand—the Westphals and the Kohns had that in common.

"Be sure to sign our book for visitors," Reverend Westphal

said, and while I did, the rest of the congregation filed past. He and I were alone in the vestibule. "How have you been?" he asked. "Everything going well?"

"Fine. I'm doing fine, although it's not the best of times for my folks. My dad's blood pressure is up. He worries about the farm, and my mother worries about him."

"Yes, I understand he's having weekly checkups. Sandra keeps me pretty well filled in." Reverend Westphal rested a hand on the doorway. "And I know that Sandra will leave a big hole behind if she and Mike go ahead with their idea to move away."

"It's pretty certain. She told my folks about it a few days ago, which went okay, though I think it'll be a while before it sinks in."

"As special as Sandra is, I'm going to miss her a lot myself, and that'll be about one-hundreth of what your folks will feel. It's never easy when children move away, and it's harder the older they are. You expect them to be around forever."

"Like the farm. I don't know how long my dad will be able to hang on. He wants us to believe he has it under control, but my mother says he tosses and turns at night, and getting him to talk about it is next to impossible," I said. "He hasn't talked to you, has he?"

"He does talk to me a little. He's very cool when he does. I think he's walking on ice. He wants to be extra careful about every step he takes. The last thing in the world he wants right now is to take the wrong step." Reverend Westphal let out a breath. He had become quiet, contemplative. "Any church meeting that your dad is at, I always know he'll speak up if there's anything that doesn't sit right. But he might go through an entire meeting and not say a word, just taking it all in, digesting things, sticking them away somewhere. Later on, it might be years later, he'll bring up something I've forgotten—he remembers it like yesterday. He's kept quiet about it till then because it wasn't important till then. That's his attitude, as I'm sure you know; he'll make his decision all in good time. I don't say he's waiting for the Lord to direct him, but he is waiting."

I nodded my head. We exchanged looks of understanding. He spoke again about patience. "One thing I've learned myself here in Beaver is to wait and listen and make sure these people with their German heritage are on your side, because once they are, they'll stay there. But don't move too fast and don't ever cross them!"

"I know," I said.

"Having gone a couple of times with your dad to the Michigan District conventions in Ann Arbor, I feel I've gotten to know him better than some of the others of his generation. He's not predictable on every issue. Yes, he's old-fashioned, no doubt about it, but, well, take women's suffrage—" Women were yet without the vote at Beaver Zion, a holdout men-only parish, one of the few left in the most tradition-locked of the Lutheran synods, the Missouri Synod. "Your dad has spoken out that women should not only have the vote but they should be allowed to be ordained as ministers. That's pretty radical." Reverend Westphal smiled what I took to be a wry smile. "On the other hand, this construction and remodeling we want to have done to the church for the centennial, I'm not sure he's in favor of that. But I'm sure he's giving it some thought." He clasped my hand again. "However things turn out with the farm, it's your dad's decision. I mean, you or your brothers don't have any plans to take it over, do you?"

....

For Sunday dinner we had stuffed pork chops, mashed potatoes, and homemade sauerkraut, cooked on the wood burners while we were at services. "What did the pastor say to you?" my father asked.

"Only that German farmers are stubborn as mules."

My father laughed forcefully, and we ate in good humor.

But after the dishes were washed, while in our car driving to visit my brother Ronald and his wife, Margo, Diana said, "Your mom was telling me about Ronald's transfer to Monitor. I didn't realize that it's been another major change for your folks. They only see Ronald about once a month now.''

Ever since Ronald was on his own, they had been able to count on seeing him every Sunday at Beaver Zion, where he was deeply involved in church affairs, but this past summer, after postponing the inevitable for years, Ronald and Margo had moved their church membership to Trinity Monitor. Their oldest child, Angela, was to begin first grade, and, because they wished for her to attend a Lutheran school and because the school bus that ran past their house delivered students to Monitor, not Beaver, they had to make the switch.

''It would be a big change,'' I said, ''now that I think about it.'' It meant a break in the automatic part of Germanic solidarity and social communion—the get-togethers of worship and church meetings, the regular events of the church calendar—and, for my father and mother, it added to their isolation from their family. ''Once a month?'' I repeated Diana's words. ''I guess Ronald and Margo have been busy getting themselves involved at Monitor.''

Their house on Eleven Mile Road was low, with white aluminum siding and cornfield borders on three sides. Ronald had planted a row of pine trees in his yard for a windbreak and screen. They were nursery seedlings, and some would be dead in spring from rabbits that squat on snowdrifts and nibble off incipient growth. There was a hunting dog with a bony fighter's face out back, but, fenced in by ten feet of chain link, he was no threat to the rabbits.

On the front-door stoop Diana and I shivered deep into our coats. Our breathing sent off clouds. ''Come in, come in,'' Ronald said to us. ''Don't just stand there.''

Indoors, even the basement was warm. To beat the price of conventional insulation, it was popular now in the Saginaw Valley to mound up dirt with a tractor and blade against house foundations. Ronald had considered it but, for the sake of the sunlight streaming through the basement windows, had lined his basement walls with Fiberglas instead. There was a pool table in the basement, and he and I began a game of eight-ball. Ronald is an overachiever in games. When he was younger he played fast-pitch softball in the Midland leagues that produced a world championship team, McArdle's Cadillac, and, recently, after taking up horseshoe pitching, he—and Margo, too—had made themselves into world-class players. In the long summer evenings of Michigan, the most western of the states in the eastern time zone, they threw tens of thousands of practice horseshoes in the pits Ronald had dug, to tournament specifications, into their backyard. Trophies won in softball, horseshoes, plus bowling and euchre, and some won with Ronald and Margo as doubles partners, enough to open a store, sat on wrought-iron shelves. A willowy athlete and a former bank teller, she had been his high school sweetheart.

Coming downstairs, Margo carried a plate of Christmas cookies. "None for me," Ronald said. At age thirty he had taken a look at himself and had seen all his after-game binging in an overflow of flesh, lower shirt buttons undone, a general image of ill-feeling looseness, and so had given up beer, sweets, second helpings. Now, with a taut belly, sideburns cut high, his sinister black mustache shaved short, he looked ten years younger, and, funny thing, looked less like a jock.

Their children, Angela and John, were chasing each other up and down the basement stairs. "You kids take it easy. We got company," Ronald said mildly. He was ignored, and he went after them: "Jeez-oh-mighty, I said stop!" They were feeling rambunctious and wanted to play. He came back with them giggling in his arms. We let them have the pool table, explaining the rules, which they made a hash of, and giving them pointers, to which they were cheerfully indifferent. This was the casual, happy-go-lucky style of competition that Ronald and I for the most part had missed.

In high school, at Bay City Central, Ronald knew the glories of winning contests sponsored by the Future Farmers of America, the FFA. Another student who performed at Ronald's level, and who could be singled out as his chief rival, was Don Rueger.

Ronald was president of their FFA chapter, Don the treasurer, and they competed in a small, regulated world of ten-acre test plots and pick-of-the-litter livestock showings. "Farming for ribbons," they called it. There are some boys, gifted with discipline, authoritatively gung ho, who have the aura of the land, and who do things—get up before dawn, read the latest literature, become impatient for the first thaw, talk crop yields to their dates—that mark them indelibly as future farmers. Years after Max Brown, the FFA faculty adviser, had Ronald and Don in his class, he said of them, "They were two of my most memorable students. If they weren't cut out to be farmers, I don't know who was." When Ronald and Don graduated, however—the moment of truth—they went separate ways: Ronald changed his mind about farming and enrolled at Ferris State to learn a machinist's trade.

"Hit the eleven," Ronald said, lining up a pool shot for his daughter. "Hit it soft." She poked with her cue stick toward the colorful sprawl, and the eleven rolled with almost zero momentum to a pocket.

"Now the thirteen," he said. But she wrinkled her nose and went for the nine. Ronald sat down.

I asked whether he thought there was any chance Don would return to work the Kohn farm next year. He said there was none. Once begun, how does a feud stop? Don and my father had seen each other at a wedding a few weeks ago but had avoided contact. Was it worth someone's effort to try to be a peacemaker? "No, I'd leave well enough alone. It could be a lot worse. They could have had a really big blowup. At least he didn't yell at Don the way he used to yell at us."

The differences between my father and Don (and Ronald) were the differences between the generation that came out of the Great Depression and the generation of the FFA that came into prominence after World War II. The FFA had educated Don and Ronald in the holy writs of modern farming—the "Commandments," as my father refers to them: Thou shalt plant rows close together (my father thought this was greedy and counterproductive); Thou shalt eliminate fences and enlarge the fields (my father's eight ten-acre fields became Don's one eighty-acre field); Thou shalt rely on pesticides as the first line of defense (my father had never used them); Thou shalt covet thy neighbor's acreage (my father had never expanded); and so forth. I had gone with my father once to inspect one of Don's cornfields. "See how thick he's planted, way too thick. Defeats his purpose. Guarantee you he'll get a lot of stalk and few cobs. And the air won't get through to dry the cobs that he does get. He'll pay thirty-five cents a bushel at the elevator to gas-dry them, and still not have top grade. Field-dried is top; it cracks into chunks. Gas-dried turns powdery. And, planting this thick, you have to pay for extra fertilizer, and you wear out your soil faster." Legitimate points every one of them, although it was also true that Don's bushels-per-acre yields were higher than my father's. Through all my father's differences with my generation of farmers—and with my generation, period—ran a single, simple objection: "You guys are in too much of a hurry to get ahead." Last summer my father had needed precision holes angled and drilled in a replacement bar for a plow hitch, easily accomplished at Helfrecht's machine shop in Saginaw, where Ronald was an assistant manager. Ronald did the drilling off-hours, and there was no charge. Then it was Don's turn to ask Ronald for a piece of machine work that could have been handled the same way, but Don was rushed and willing to pay a surcharge. Because of his hundreds of acres, he had to be as quick as possible. He planted in "one swipe of the field," slapping in seed, fertilizer, and pesticides, and at harvest, he ran his combine through in a high gear, bouncing kernels onto the ground. My father would salvage what he could of Don's spills with a shovel and bushel baskets after Don had moved on to another field. No wonder their partnership had not lasted.

"There are other guys besides Don. He can find someone new to cropshare with," Ronald said to me.

"Do you think he'll do that?"

"What's his choice? He can't do it by himself."

Margo forced another cookie on me. ''How is Dad doing?'' Her soft voice mingled with the carols on the stereo. The music was loud, and she interrupted herself to lower the sound, walking past the trophies swallowed up by the colors blinking from a Christmas tree. The music faded out, and Margo returned. ''We've had so much going on we haven't seen him and Mom in I don't know when.''

''He was going full tilt when we were here in November, cutting wood, shoveling snow, grinding chicken feed,'' I said. ''This time he's staying in the house a lot. He goes to bed at the regular time, but sleep doesn't come, and he's up before it's light.''

''Like always,'' Ronald said.

Diana, looking warily out the casements at windblown snow, said, ''I think the cold is keeping him inside. It's miserable. I can't remember it ever being this cold.''

''True enough,'' I said. ''Anybody tromping around in this weather is asking for a heart attack.''

Margo sighed. ''This winter already feels like it's been here forever. I can't believe we still have four months to go.''

....

In the morning the driveway was gone, and the road, and half the height of the chicken coop. I rubbed my eyes awake and thought of Times Square as I used to draw it with crayon, the glossy skyscrapers and fantastical streets of a boy's imagination, until on a college-impulse weekend, there in front of me Times Square stood, dragged down from my enchanted presumption to the sorriness of a panhandling wino. I thought, too, of a college friend who, on his first visit to a farm, had much the same disappointment—''Some cow slobbered on me, and I got mud and shit all over my pantlegs.'' This morning he and I could have been boys again. The world outside, a piling up of deep, white snow, was at once tranquil as his imagined countryside and brilliant as my cityscape. It was the Great White Way, a place where one could walk dreams.

Out the window I saw a pickup truck and an American-made car come twisting down Carter Road from the north. As they passed the house they lost momentum, and, tooting and spinning, they mired down together astride a great snowbank, long and flat and deceptively high, in a spot just south of the house, a spot of historic snowbanks that had imprisoned horse-drawn carriages, Model As, yellow school buses. Modern as Carter Road was now, it was not, in this spot, weatherproof. Ahead of the truck, my father already was shoveling a single track through the snow. I hurried out and began making inroads on the opposite side of Carter. The driver of the truck, a middle-aged man, gunned it, seeking purchase on the asphalt. "Take it easy," my father shouted above the screech of the tires. "We'll get you out." The tires hit asphalt, and the truck lurched forward, stopped, and the tires screeched some more. My father shook his head. The driver of the car, a young, pallid-looking woman in a quilted polyester coat, came forward and told us the man in the truck was her father. He had been trying to blaze a path so she in her car could reach a main road, one that had an advance guard of snowplows. "I work at Dow," she said, "and I'm due in at nine." She and her father were newcomers to Beaver Township. This was their first winter. Sheepishly, her father climbed down from his truck and joined us with a shovel he had had the foresight to bring along.

The sag in my father's spine was made conspicuous by the shoveling. From time to time he straightened up and arched his back to get relief. With his short, economical strokes, though, his progress was faster than mine. Finished with his side of the road, he moved over to assist me and the driver of the truck, who had a lunging stroke that was virtually useless; the man's shovelfuls weighed more than he could lift. He began to breathe heavily, and his daughter put an arm around him to slow him down. Seeing this, my father called a halt. We all rested on our shovels.

We were standing in snow up to our waists. "Nothing compared to the snow we had in 1936," my father said. The snow of that winter, the worst in the memory of Beaver old-timers, was, in several places, above a man's head. The township was snowbound for days. A visitor from Detroit, tired of the wait, found he could walk on top of the snow and tried to get out along Carter Road. Reaching a snowdrift in this very spot, he fell through. My father found him sometime later, and carried him like a feedsack to the nearest phone, at Albert Pashak's place a quarter mile south; but the man was dead.

After our rest, the shoveling went easier. The driver of the truck climbed back in, revved up and successfully followed our dug-out tracks. His daughter, searching through her pockets, offered my father a tip. "Thank you all the same," he said. "Glad to be of help." She expressed surprise. Smiling, she left in her car, sashaying in and out of the tracks but making it through.

My father and I walked to the house across the cover of snow crunching underneath, and relaxed by the kitchen stove. The painful tingle of warming up felt good. In the excitement, I had not realized the temperature outside was at zero. My father closed his eyes. That he would spend an hour shoveling snow was the routine of his winter, and probably I would have thought nothing of it if I had not remembered my own burst of hard reasoning from yesterday that this was heart-attack weather. Work made my father feel better, as my mother had said, but strenuous work risked his heart. That was the catch. And there was something else. I had read about the recent deaths of two farmers. One had let himself fall into his combine, or so it appeared; another had drowned in his livestock pond, and it appeared he did not try to save himself. These may have been accidents, although their neighbors did not think so. A friend of mine, Paul Hendrickson, having investigated several other deaths in farm country, many obviously suicides, wrote in the *Washington Post,* "One would be a fool not to think that something is terribly awry in the middle of the country." Ronald had told me that a farmer he knew, a man our father's age, had not been outdoors in weeks. He would not get out of his pajamas and lacked an appetite. His family had hidden away his guns and tried not to leave him alone. Clearly he had lost his will to work, perhaps his will to live. My father in his kitchen chair was flushed from work. But did he have some half-conscious nightmarish motive as he energetically cleared snow from the road? Or was I scaring myself? I did not know. I had read that German farmers, formidable in a storm or in an argument, are the least able to bear up if they believe a disaster is of their making. Writing in *Psychology Today,* Val Farmer, a psychologist who has counseled desperate farmers, concludes that "the Germanic families . . . are having the more difficult time of it. Their farming strategy is more emotionally driven, fueled by persistence, hard work, commitment, conservative fiscal management and close family participation. When things fall apart for these families, an entire value system is jeopardized; the emotional cost can be devastating."

I might have been a better son, I thought, if I had not had to deal with so many intangibles.

....

The way my father had defined things, he had three alternatives: enlist someone else as a cropsharer; risk his health and go it alone; or sell the land and retire. That night, expecting sleep, I got instead a wild thought. There was a fourth alternative.

In the morning the thought hung around. Diana and I had no commitments, none anyway that couldn't be put off or canceled, no home (a suburban house outside Washington, but so what?) and no children (therefore no home). From my first marriage, I had a daughter, Liz, who lived with her mother and stepfather near the Michigan State University campus, and that was part of what I was thinking. To be closer to Liz, Diana and I had already moved from San Francisco to Washington, and, by that logic, we could move to Beaver Township. Not forever. One year, I thought. I did not know.

"Don't say anything until you think it through," Diana said.

I went for a walk. A big devouring sludge-gray sky claimed the smoke from the chimney. I tracked over my father's frozen footprints across new snow from the house to the barn to the chicken coop. I watched him fill wooden troughs with ground-up oats and corn. The chickens flapped and ran about, as if from a raid, raising the dry manure. I took down an egg pail from a shelf and began to reach under nesting chickens for eggs. The nests were tall shallow boxes resting on short stilts. They were motley—some of wood with square holes, eaten by termites, so wasted they could be smashed up for kindling with bare hands; some a setup of tin boxes with round holes, showing rust. Structurally the tin nests needed more bottom weight. One had toppled over last year and killed a California brown.

The metal radiated cold. So did the mossy, wood-shingled roof. Heat was on the floor, forget the absolutism that ceilings are always warmest. The floor was a foot below a packing of straw and manure. I could feel the warmth of the manure through my boots. The chickens looked healthier today, except for one tottering about, leaning on a wing to keep from falling. She sat down with sick eyes, her head swiveling, feathers lusterless, the shadow of death on her.

I walked out of the coop and into a cornfield, searching for a trace of the old cow lane. The wind was keen, and I kept my head down. It may have been the wind, the hurting cold, but I began to doubt that there was anything I or anyone else could do for my father and his farm at this late date.

The Chippewa chief Sassaba once said about Michigan:

> *Beware the dead time of winter when no living creature stirs and great doubt sits on the land. What cries come from the ground. Beware and walk lightly on the fathers of this land, on their bones. Do you not see how they are conscious of our step?*

In the woods the other day I had been conscious of their step, of the step of Heinrich Kohn, the first Kohn to come into the

Saginaw Valley. And what had he been conscious of? Any special sense of this place? Any heartfelt something? Had he in his wildest-eyed, farthest-fetched imaginings ever thought of himself as a father of this land?

Another question: Can land that you lose haunt you like lost souls?

Heinrich had recurrent fears about a snowstorm that would blow in and destroy the farm. Storms of tremendous power are not unknown in Michigan. The first Kohn house was made of heterogeneous planks and rafters, an impermanent, thrown-up frame box. Heinrich replaced it with a tall, prideful fortification of brick. The barn he put up was massive, three hundred feet long, seventy feet wide, forty feet high, one of the largest in the township, with thick wooden walls and floors solidly anchored by wood and cement. He set his corn cribs on cement pillars. His pigpen had stone walls. Somewhat unusual for their time, nearly all his buildings had cement bases, meant to survive any storm in a doubtful winter. "Beware . . ." Chief Sassaba had said.

Beware of a doubtful future, he might better have said.

....

Heinrich Kohn was born in Germany in 1841, a younger son on a farm that was no longer divisible, and he had to make his living at first as a mason in Berlin, laying brick and cobblestone. At the Berlin Zoological Center he was hired for a job of walkways and walls. Once up, the walls had at each gate a square of wood for posting circulars, and it may well be that Heinrich first became aware of the Saginaw Valley by reading circulars sent to Germany by the state of Michigan. OPPORTUNITIES! UNDISCOVERED RICHES! FERTILE SOIL! A MILD CLIMATE! The Saginaw Valley was advertised as a sunny location on the same latitude as Marseilles and Florence, which is correct but misleading, and there were other exaggerations. The picture of the Saginaw Valley that the circulars gave was not the brute American frontier depicted in European sketches and wood carvings, the America of wary, resentful natives known fearfully as "wild Indians." That, of course, was the America of Columbus, nearly four hundred years in the past, an America that to urban Europeans already seemed quaint. On Heinrich's arrival in the Saginaw Valley, then, he may have been shocked. It must have felt like stepping back to that past, into a scene of "early savage America." From his wagon he would have seen tall, rangy forests, swamps with tussocks of coarse sedge to twist an ankle, more forests, more swamps, and glimpses of Chippewas and Potawatomies in bark hunting lodges. Except for the Hopewell People, who grew squash and beans in riverbank gardens in the period around 1000 A.D., no one had ever tried to live in the Saginaw Valley by planting things. Heinrich might have seen the beginnings of European-style farms—a few emigrants had started to clear land, bringing grain seeds with them in pig bladders—but, on first glimpse, the valley must have looked like one of these early European sketches. And in America, it had a reputation as wilderness. The U.S. surveyor general, Edward Tiffin, had sent an expedition of surveyors into the Michigan interior, and, from their eyewitness report in 1815, he had written that it was truly forbidding: "Not more than one acre in a hundred, and, were the truth revealed, not more than one acre of a thousand, will admit of cultivation. Nothing but Indians, muskrats, snakes, bullfrogs, and mosquitoes exist, and there appears to be no incentive for white settlers to move in."

Morse's *Geography* quoted this account, and it was taught widely in American schools. In 1831, Alexis de Tocqueville traveled into Michigan as far as Pontiac, which is today a satellite of metropolitan Detroit. The Saginaw Valley is about one hundred miles north and west, in the flood plain of a Lake Huron bay, Saginaw Bay, and de Tocqueville thought he and a companion might go there. An innkeeper could not believe it:

> *At the name of Saginaw, a remarkable change came over his features. It seemed as if he had been suddenly snatched from real life and transported to a land of wonders. His*

eyes dilated, his mouth fell open, and most complete astonishment pervaded his countenance. "You want to go to Saginaw!" exclaimed he. "To Saginaw Bay! Two foreign gentlemen, two rational gentlemen, want to go to Saginaw Bay! It is scarcely credible . . . Do you know the forest is full of Indians and mosquitoes; that you must sleep under damp trees? Have you thought about the fever? Will you be able to get on in the wilderness and to find your way in the labyrinth of our forests?"

Michigan was barely part of America. To bring white people and roads and commerce to the empty middle of the state, to exploit its resources, to imitate the Eastern seaboard, official Michigan came up with a marvel of a promotional campaign. Into it went all the enterprise of America, all her genius, and all the credulous wonder of Middle Europe. Not the least of the credit should go to Max Allardt, a Michigan citizen born in Germany. In 1869, the Michigan legislature authorized a special envoy's status for Allardt. He returned to Germany with cardboard circulars (MILD CLIMATE!) and copies of a pamphlet, "Michigan, Seine Vorzuge Und Hulfsquellen, und Vollständiger Karte des Staates" (Michigan, Its Advantages and Resources with a Complete Map of the State). In 1881, Allardt's successor had the pamphlet updated and translated into Dutch, French, and Swedish. Additional agents were paid by the state to recruit in the city of New York among solitary, strong-looking emigrants. Then, in 1885, nine years after Heinrich's arrival, the state recruitment office was abolished. Allardt's pamphlet was discontinued, and in 1893 it was disclaimed. The Michigan legislature issued a new pamphlet, "Michigan and Its Resources": "Emigrants who float in with the tide will not find Michigan their El Dorado. . . . The Lord hates a lazy man, and the law takes care of the dishonest. Michigan is a hive of industry, with no use for drones."

"A hive of industry" is a good opening line for a book about the Saginaw Valley at the end of the nineteenth century. In twenty-five years everything had changed. In Bay City, at the center of the valley, the millionaires per capita—58 among 30,000 — outclassed Manhattan and outclassed San Francisco. UNDISCOVERED RICHES! The circulars, incredibly, had been true. The treasure of the Saginaw Valley, which the U.S. surveyor general's expedition had missed, was its oak, hemlock, maple, and the incomparable white pine. In California, in gold country, miners had struck it rich, but not as rich, according to the Michigan Chamber of Commerce, as had the Saginaw Valley lumbermen. The Saginaw Valley timber supposedly earned its finders $1 billion more than the California mother lode. On the Saginaw River there were 112 sawmills, a jammed-up, noisy assembly line that cut 23 billion board feet. In sheer volume, the *American Lumberjack* magazine reported, "no other single lumber producing area in the world can equal this."

Until Heinrich married Augusta he lived with other sawmill workers in a big wooden house, a setting like a commune. He had an unmarked face ("smooth as an Indian's" it was said), a small hook to his nose, a high forehead, eyes that were deep-set and dark and fierce. He had a philosophy—"The Lord hates a lazy man"—that was the official state philosophy. His job was six days a week, and it paid $1.50 a day. On Sunday he attended German Lutheran services. As far as it's known, he possessed none of the standard vices in a city whose most active street was Hell's Alley. Friends who partied must have liked to have him come along as a sober bodyguard. Most of his pay he saved for the land he purchased in 1879 in Beaver Township, twenty miles outside the city. Change was on a slower course in Beaver Township. Fifty years after Heinrich planted his first crop, a significant portion of his farm was yet in trees, and a Chippewa family lived in a peeled-bark teepee a quarter mile south, at the corner of Carter and Seidlers roads. My father—Heinrich's grandson—passed by it walking to and from the school at Beaver Zion.

....

We were at the kitchen table. Supper was almost over. "You could get a small fortune for the barn if you broke it down and took it to New York," I said to my father. "People love the old boards for wall paneling."

"How could they be worth so much?" my mother asked.

"In New York, everything is worth a lot," I said.

"In New York, everything costs a lot," my father said, forcing the distinction.

He had seen New York only on postcards and on television. A tall, sharp cleanness, an imagined cleanness, was transmitted by the flat, far-off pictures of the big city. The barn boards he had seen up close, with their rough feel and unmistakable smell—"Who in their right mind would want those dirty old things indoors?"

But he was in a good humor again, and he had a reason for it. "I spoke with Gerry Radke, and we agreed that he'd cropshare the eighty next year," he said contentedly, stroking his chin. "A one-year deal." Gerry Radke was the youngest son of a neighboring farmer. He was only twenty-one but knowledgeable about land. He had shy, finely chiseled, hemp-colored features that looked at home under the visor of a farmer's cap. More to the point, he probably was better suited to my father than Don or any of the Kohn sons. Gerry had an old McCormick tractor with an umbrella propped in the wheel hub, and he did not have a conglomerate of farms to keep track of or a separate life to be given up.

"He thinks like a farmer," my father said.

"I guess that means he thinks like you," my mother noted with a subtle tang.

"He's so young he'll be almost like a grandson," was Sandra's assessment later that evening. "And when Daddy is being Grandpa he's a different person. He's not so hard. He doesn't take any guff from my boys when they're out working with him, but they don't have to be perfect like we had to be. He lets them off the hook."

"He mentioned just yesterday how much he's going to miss your boys," I said to Sandra.

"Your mom told me the same thing," Diana said.

"Have they said anything else about us leaving?" Sandra asked.

"No," I said. "I thought they'd be more upset. I thought they'd come up with a thousand reasons why you shouldn't go."

Sandra was beating eggs in a bowl with a spoon, and the patter grew louder. "You know what?" she said. "The last few days I've started to have second thoughts. Everywhere I go now I think, 'When I come for Christmas in a few years, I'll be a stranger here.' " She wiped the back of her hand roughly across the corner of her eyes. "I promised myself I wouldn't cry when I told Mother and Daddy, and I didn't. I was cool, calm, and collected. But I had the butterflies for a whole week before. It was like—I don't know how to explain it—it was the hardest thing I've ever done. I sat in the kitchen and told them, 'We're not leaving tomorrow. It's not like that. We're waiting till Mike graduates, but I wanted you to know ahead of time. Our minds are made up. We're definitely leaving.' "

We heard David banging through the back door, followed by Mike, who had collected him from basketball practice at Beaver Zion. David was of precocious size and much sought after by the Zion coaches. Had Mike, with his size, wanted the punishment, he himself could have hooked on with a team at Ferris State, but he hadn't. There was a sense of the grownup in Mike, and part of what he had outgrown was being an amateur sports star. He enjoyed being a fan, most of all for Sandra, who played slo-pitch softball in a Midland women's league. Of the husbands and boyfriends at the games, Mike was wholly atypical. "A guy who yells for his wife but never yells at her" was a characterization of Mike you heard from Sandra's teammates.

"Hi, honey," Mike said, waving at the kitchen door. "Baking more goodies, I see."

"What else?" Sandra's smile shone over the tears that had been there.

He came over and stood beside her. There was a round metal engraving of a sailing ship on the wall, and below it the two of them conjured a portrait of newlyweds off to see the sights, feeling alive and worldly.

....

"Yep, like my grandpa used to say, we live on the sunny side of the road," my father had said, after making his arrangement with Gerry Radke, and I was glad. The last month had produced all sorts of tangled-together emotions, old confusions, old resentments, new hopes in with new fears, and, most of all, a futile, inexpressible wish to come home. But I had not been called on to make any hard decisions. I had made no decisions at all. I had escaped.

In a sense, my father had also, although I did not learn until later that for two weeks he had kept a bank voucher for twelve thousand dollars in his desk—option money thrust upon him by the lease hound after we left in November. Now my father gave back the voucher, uncashed.

Almost as soon as he did, other lease hounds came to the door, one from Oklahoma who said he represented a company called EMCO, another from a Texas company called Cochran, and another from Texas, who identified himself as an agent of the Crockett Oil Company. The Crockett man was about thirty, well dressed, and he had the confident, earnest geniality of a serious salesman. "By spring we're going to be the only company doing business here, or, let me put it this way, the only one worth doing business with," he said. He was prepared to outbid the competitors and buy out any leases they held. "What's the best offer you've had?" he asked my father. "Whatever it is, I'll do better."

"I'm afraid I'm not interested," my father said.

"Okay, you don't have to tell me. I'll give you a hundred and fifty. I know no one is beating that." Multiplied out, that was an offer of eighteen thousand dollars for my father's drilling rights.

It was a remarkable number. Last summer the typical offer was ten dollars an acre. By hemming and hawing, a farmer might have raised it to twenty. Now, after drillers had hit on several holes in succession, the Saginaw Valley oil rush had taken on a reckless, headlong mood more in line with the gigantic reservoirs of Alaska or the North Sea. It did not seem to matter that the pool of oil beneath the valley was inconsequential by comparison, or that it would be played out within a few years, or that these oil companies, know as ''strippers,'' were on the move from one geological formation to another across the Farm Belt. What seemed to matter to them was their speed and thoroughness in overcoming all obstacles and claiming the land for themselves.

Apparently the Crockett man had been briefed about my father. ''You may have heard some stories about the damage these rigs do to a farm,'' he said, ''but I wouldn't believe half of what you hear.''

''Well, since you brought it up, let me ask you a question,'' my father said. ''We've got a couple of problems around here that we didn't have before the drilling got going. For one thing, our well is down next to the bottom''—Ronald had told me, ''You should have been here when we dug out the well,'' as if I had missed a big horseshoe tournament, explaining how they had crowbarred off the well cover, shinnied down a rope into the dark hole, and filled dozens of buckets of gray, mucky clay, pulling them up by the rope—''and, another thing, the water we do have is changed and looks rusty now, almost scummy, and doesn't smell right.''

''You think it's because of the drilling? I can't see how that's possible.''

''Well, I know they're using a lot of water to prime these wells, and they're drilling holes right through the water table. And I know they use chemicals when they drill, and they use solvents to break up the viscosity of the oil, what they call 'acidizing' it. So where do those chemicals end up? And the brine and the minerals they bring up from down below—where do they end up?''

The Crockett man whistled his surprise, a little unwillingly. He ran his fingers through his hair and lifted one leg onto his toes.

"You know more than most folks I've talked to, I'll grant you that. But with all the environmental regulations, Uncle Sam is watching everybody. If there was any groundwater contamination, the federal inspectors would know about it."

"You must have different ones in the oil business than we've got in farming. The big grain companies are dumping dirt and chaff into the grain the U.S. ships overseas, and the federal inspectors don't see a thing," my father said soberly.

The Crockett man smiled—he had a fabulous smile—and reintroduced the subject of money. "You'll be getting five to ten times more than some of your neighbors got," he said.

Nothing in my father's face changed, nothing. "Turns out to be a pretty good trick on them, doesn't it?" he said.

"If you don't tell them, I'm not going to. Anyway, once they've signed, that's it."

My father pulled a work coat and sweat-stained cap off a clothes peg and put them on with an urgent air. "Sorry to disappoint you," he said. "But I'm not interested, like I said."

"Think it over. I'll be back." The man replaced his hat and left.

Minutes later my father left also in his pickup, giving my mother a ride to Beaver Zion. In the church basement she met once a week with Reverend Westphal's wife, Carole, and eight or nine other Zion women—the Quilting Club—who had obligated themselves to make seventy-five or more farm blankets for Lutheran World Relief. Each had a bottom sheet of white cotton and a quilted top of squares and triangles cut from charity boxes of old, gaudy garments. Today, one was going to have an unusual centerpiece, an oval graphic of yellow chicks, from a 1940s cotton feedsack my mother had found upstairs.

While she quilted, my father returned and sat with Diana and me to eat the beef patties left simmering on the Kalamazoo. He refused seconds. "Nope, didn't do any real work yet today." A semi-truck with a long flatbed, brake lights flashing, drove by and turned into the Radke place, owned by Gerry Radke's father. A pile of steel trusses was strapped to the flatbed. A successful hole had been drilled last week on Radke land, a quarter mile north and across the road in Section 20. ''I watched them put the rig up. It's portable, and they bolt it together,'' my father said. He pulled aside the frilly kitchen curtains for a better look at the flatbed. ''Looks like it's carrying a rig. Must be going to drill a second hole.'' A crane began unloading the trusses like heavy-gauge Erector toys. At the first hole a grasshopper pump jumped up and down in place like another adult toy. ''They hit oil, but so far, from what I understand, there's more brine than oil. Guess they figure a second hole will do better.'' He stood at the window, frowning, stretching his fingers, flattening his palms. ''Hear that pump? *Chuk-ka, chuk-ka.* Kept me awake half the night. That's another reason I'm not signing. These lease guys tell you they're prohibited from drilling too close to your house, but I've read the contract. They can drill eighty feet outside our bedroom. That'd be something. All I'd hear day and night is *chuk-ka, chuk-ka.* Enough to drive a man nuts.''

The postwoman in her car paused out front with a delivery of Christmas cards and advertising sheets. I threw a coat on and ran to the mailbox. *Chuk-ka, chuk-ka, chuk-ka,* the rhythm of the pump across the road was repeating like a stutter.

Later in the afternoon, home from the quilting bee, my mother blurted out, ''Striking oil! That's the main thing we talked about today. Who's going to be the next one to get rich?''

....

The next morning, while my father and I were grinding chicken feed in the barn, a teenage boy, gangly, a lost-looking kid, stumbled up through the cow manger toward the racket of the hammer mill. He was going from farm to farm, he said, to find drinking wells with ''something odd about the water.'' My father brought him up to the house. ''Taste this,'' my father said. The boy, making his survey, probably drank four or five gallons of water a day. He swilled the water in his mouth and swallowed, screwing up his face. ''Real bitter,'' he said. ''We can do a test for you. No charge to you.''

''Nothing in life is free,'' my father said.

''Honest, our test is free.''

''Then you must be selling something else.''

The boy was not highly experienced at sales, and someone must have told him not to reveal his product until after he had run his test on a well. Holding and sipping at the glass in his hand, he continued to complain about its poor quality. He was evasive and was not making a good impression. But the water did taste bad, and it carried a faint sulfurous whiff of minerals common to depths, It had even become grainy; particulate chunks were visible to the eye.

''Go ahead and test it,'' my father said.

''All right, good,'' the boy said. ''I'll take a sample, and my father will bring back the results. Probably tomorrow. He's usually one day behind me.''

THREE

Much that is underfoot in the Saginaw Valley is the stuff of prosperity. The soil is a sandy alkaline loam. White navy beans grow better here than anywhere else, and, of the total world crop, a majority is from this valley. The souvenir license plate on my father's pickup reads BEAN COUNTRY, U.S.A. In England, where Diana and I have spent time, we ate side dishes of Saginaw Valley beans for breakfast. Sugar beets grow fat as soccer balls. The soybeans are superior, as are pinto beans, Mexican black beans, corn, wheat, oats, rye, barley, potatoes, sweet clover, and alfalfa. The hay is aromatic, and we used to take three cuttings a season. And beneath the rich soil are more riches, the remains of swamps over which glacial ledge rock drifted and pressed out bituminous coal, limestone, oil, and gas. From an ancient lake that was without an outlet are left layers of salt and, in porous rock, great cisterns of brine. The brine holds magnesium, chlorine, calcium, sodium, and bromine.

A sense of this potential wealth—and something of a boom philosophy—informs the valley cities of Bay City and Midland. Even while the valley was being lumbered out, smaller booms asserted themselves. Coal mining and manufacturing and farming, overlapping one upon another. Bay City and Midland spring very much from the late nineteenth century and possess a genus loci: points on the map, equidistant east and west from the Kohn farm, where entrepreneurial forces met to form the two cities, twin ideals of early urban America. Bay City and Midland are inconceivable outside the United States, so archtypically American are they.

Bay City tells its early history neatly in an inventory of its most successful companies: the Michigan Pipe Company, exporting wooden industrial-strength pipes around the world (in many European cities, drinking water flowed in Bay City's wooden pipes); the Bay City Salt Manufacturing Company, burning wood scraps to distill salt from brine; the Bay City Potash Company, making soap from wood ashes; the Monitor Coal Mining Company, shoring up unstable, waterlogged mines with oak timbers; the Aetna Portland Cement Company, burning coal to turn limestone to cement; the German American Sugar Company, refining sugar beets into granulated sugar and molasses; the Michigan Chemical Company, turning molasses to alcohol. Of these, however, the only two to prosper into the 1980s were the cement factory and the old German American sugar beet factory, where, off and on, my father moonlighted, and which had been bailed out by investors from Saudi Arabia. At present, Bay City was between booms. Lesser gods had returned. The arching middle span of the Third Street Bridge, one of three for carrying Bay City traffic across the Saginaw River, had fallen into the water one day in 1979, where it could be seen for years rusting, a snag for river litter, its repair too expensive for the city budget.

In Midland, on the other hand, the rise and fall of the lumber industry had been followed by the rise and continuing rise of the Dow Chemical Company. Herbert Dow, the company founder, a chemist, had recognized in the mineral contents of brine the pesticides and plastics and scores of other products that could be made. Dow Chemical rose above the booms into a superlative, mythical realm—and took with it almost everyone living in Midland. In the 1930s, headline writers admired Midland: THE TOWN THE DEPRESSION FORGOT. If true then, it was not quite true currently. As hard up as factory workers and farmers were in the rest of the valley, Midland could not help but feel some of the changing times.

And yet a new boom was in the making, and its makers, Crockett Oil, EMCO, Sun Oil and the rest, exploring ever deeper below the valley surface, had attracted a kind of camp follower. This was the water tester.

The man who tested the well water at the Kohn farm did not say, at least not directly, that the results—high levels of magnesium, bromine, and other minerals—were attributable to oil exploration. He did say, though, that oil drilling can diminish the water table, increasing the likelihood that water flowing into a well will be heavier with minerals. "Which is what's showing up in my tests all through this area," he said.

"Can a person get sick from drinking this?" my mother asked.

"No, but your teeth might fall out." The man laughed, then recovered quickly. "No, it's not poisonous, not that I'm aware of. But it sure doesn't taste good. And the minerals will leave stains."

My mother nodded in agreement. "The bottom of my steam iron gets gummed up with brown junk," she said. "It's got so I'm constantly scraping it off, or there'll be brown spots on everything I iron. And the bottom of my kettle is even worse—big flakes like pieces of peanut brittle."

"It may clear up and return to normal, but my experience is that it won't." The man glided into his sales pitch. He had the sureness and humor his son did not. The free test, the understated test results, his timing, his rebound from a failed joke, the patience for the right cue, all bespoke much practice. His was an art form straight from the late-night shows. He opened a brochure onto the kitchen table. It explained the mechanics of the water-refining machine he was selling. "Our machine runs your water through fifty pounds of salt, which is a natural cleansing agent," he said. "It will remove minerals and the mineral taste. You'll notice the difference right away."

"That would be a blessing," my mother said.

"Okay, now for the sixty-four-dollar question: How much?" my father said.

"Fifteen hundred, installed."

My father considered it. His hands were sunk into his pockets.

Shadows came into his face. No appliance or piece of furniture in the house cost half that much. "That's a lot of money," my mother said, sobered.

"There's a manufacturer's guarantee. A lifetime guarantee, for as long as either of you are alive," the salesman said.

My father held out a hand. "I'll go to the bank and get you a cashier's check."

But the installers who hooked up the filtering system the next day did not have the guarantee with them. They looked blank.

"Unless I get a guarantee on paper, I don't want this thing," my father said. "Better take it back with you."

"We can't do that." The installers left it in the basement.

My father's face was hot, almost on fire. He tried to phone the salesman, who was out on his rounds, and dialed again and again until the salesman answered, late in the afternoon. His day had not gone well either. Confrontationally he demanded to know if he was being called a liar.

"Call it what you will," my father said.

The salesman's car was in the driveway early the following morning. "Here is your guarantee," he said. It was handwritten. "This is my personal guarantee. Better than what any company will give you." At the top he had printed with a black pen, LIFETIME GUARANTEE. He was apologetic. "The company's guarantee is five-and-dime. This one is worth something." My father examined the saleman's sheet of paper and, I guess, examined the man.

"Okay," my father said with some reluctance. "A deal is a deal."

My mother thanked the man. "Our water *does* taste better."

....

After he left, my father and I split firewood. The cut-off trees we had left on the back forty in November were now a huddle of blocks next to the barn. They were barely visible through the snow covering.

I picked out the heaviest ax and received familiar advice. "Don't swing so hard. It's in the wrist. Like hitting a baseball." My father had been an ageless ball player, at fifty still the catcher on the church team.

"I always tried to overpower the ball," I said, a habit of impertinence years old. But I corrected my swing so the ax blade twisted fifteen or twenty degrees at the first bite of wood. "Damn cold out here," I said. My father looked at me with a look of appraisal. He shrugged and swung his ax. A layer of blue sky flooded at the snow line.

Endeavoring to see myself as perhaps he did, I saw a man who had gone pell-mell for years and was fast approaching middle age, unreconciled to the place and family he once belonged to.

"Yeah," my father said. "It's cold, too cold to be out here." We struggled with a few more blocks, then went in.

I found Diana upstairs, changing the sheets in the bed we slept in.

"Sometimes," I said, "I wonder what I'm doing here."

"You're helping your dad and waiting for everybody else to show up for Christmas," she said tonelessly.

"I don't mean that—I mean me! What am I doing here?"

Diana came jerkily across the room and encircled me with slender arms, the veins on the underside striped blue. "Are you upset about Gerry Radke?"

"No, I'm relieved. What makes you ask that?"

"I know you. Once you have a scenario in your head, you hate to let it go." The words threaded between a reproof and a comforting family joke. "Remember?" she said. "That's how you got me."

"A good thing for you that I'm bullheaded," I said. I was married to Kathy when Diana and I fell in love, and, scandalized by our feelings, Diana moved to a town fifty miles away, denying me visits for most of a year. All the things we had been sharing—long Sunday rides, political activism (she was on the central committee of the Human Rights Party, a progressive third-party movement), sports (she had been a sportswriter in high school and college), the rambling talks (she was also a journalist and a historian)—were reduced to a heartbeat of letters we sent each other that year: she insisted we were entitled to no more. And when we did marry, untraditionally, before a San Francisco fireplace, without our parents, Diana in old lace, I in an Afghan vest, she worried because a week later my father and mother were to show up. "How can they not like such an old-fashioned girl?" I teased, and, indeed, they had been won over.

Squeezing, Diana kept her arms around me. "I feel that you want to get involved somehow in the farm," she said.

"My only concern here is the future. Things cannot go on the way they are indefinitely."

"But it's settled for now."

"Yes," I said, too aggressively. Diana's arms grew nervous and uncertain. I felt a temptation to unload in unreasonableness, but I shut up, and she kissed me and let me be.

....

Minutes later, Diana called up the stairs to me. "Your dad is going to the rest home to arrange for your grandma to spend Christmas with us. Why don't we go along?"

There was no good excuse not to.

My grandmother, Johanna Kohn, was ninety-two and crippled, but radiantly alive in her spare, fluorescent-lit room at the Colonial Rest Home on Bay-Midland Road. For much of the time I was growing up, she and my grandfather, Johann, lived in the farmhouse, staying until it held ten of us. When Ronald and I were very young, she would play with us in the haymow, chasing us around with a type of exuberant mischief missing from the other adults we knew, and in all my memories of her she is in love with life. A child's memory, it is said, is open for a second and then snaps shut and rolls forward, blank and flat, preserving snapshots, but, perhaps because Johanna was so different, my memories of her are full, large pictures. One is from a winter afternoon up in the attic, when I was seven or eight. She and I were sorting through a box of old clothes to find fabrics she could braid into oval rugs. "How about this?" I said, and pulled from the box a reddish linen blouse with a fragrance distinct from the attic, a young woman's perfume. Johanna took it and let it fall gently across her knees. "There should be a black skirt," she said, and there was, long and slender, with rosette buttons at the waist. She stood up, smoothing it with a brown hand, and held it against her. "Might still fit," she said. Something bittersweet, I think, was awakened in her. She seemed to be remembering herself in the blouse and skirt, a beautiful country girl, her face narrow and highboned, skin nut-brown, smelling young, a kiss-me curl down her forehead, long, fine hair loosely knotted at the back of her neck, her life yet to be. She could marry any of a dozen boys—why not Henry Kohn, the youngest son of Heinrich? Henry: an impulsive, wavy-haired merrymaker who, in his Model A, took her for Sunday rides and who clowned flirtatiously with her for a camera at the Wenonah Beach amusement park. Henry: a farmer's son who did not like farming and who, with his degree from the Bay City Business College, could have put her in a house with sidewalks outside. If in her heart she had stepped across the gone decades, and if she was wishing for a different life, with a new outfit every Easter, how could I blame her? Four or five minutes passed, a long time, while she examined the red blouse and black skirt, and I poked elsewhere in the attic's hodgepodge. Old calendars covered jagged lines in the plastered walls. Pictures under cracked glass were in stacks next to my father's schoolbooks. I found a jack-in-the-box that sprang up a mocking head tilted to one side where the neck cloth was worn through. Johanna laughed. She pulled her bun of gray hair tighter. She was done with looking at her past or, as I supposed, at the caricatures of might-have-beens. She was looking straight at me with clear, calm, shining, indefatigable, contented, tearless eyes under an old brow. "My, my," she said. "See how foolish you get when you're old, acting like you're sixteen again. I better put this away before Pa comes up here and sees me." Pa was Johann, an older son of Heinrich.

Johann: a thoughtful, serious man who liked farming and who had married Johanna on April 22, 1914.

In 1972, then a widow living by herself in a little house we had built on the back forty, Johanna broke her hip in a fall down the basement stairs. Her upper leg was bent horribly. She could not crawl. The phone was upstairs anyway. She lay on the cement floor an undetermined time. Upheaving frost had cracked the cinder block foundation, letting in a thin seep of water, which she licked off her fingertips. She stayed awake as long as she could in her vigilance for my father's knock—sooner or later he would make one of his regular visits or would miss seeing the blue wisps of her chimney smoke—but eventually she drifted off. ("I thought, when I wake up I'll see Pa"—who was in heaven.) Found in time by my father, Johanna awoke in Bay City General Hospital. Her hip was rebroken so it could be properly set. Discharged, she walked with a four-legged metal crutch like a child's training aid. A year later, when she slipped and again fractured her hip, she had to give up the crutch for a wheelchair and had to move into the Colonial Rest Home.

Three years had passed since I last visited her there, a lapse that disturbed Diana, for she had taken to my grandmother even in her broken, doll-like condition. "Authors who write about downtrodden farm women ought to meet her," Diana had said. "You can feel the life in her. She lights up a room with it." Yet, when I had opportunities to see Johanna, I began to make excuses. Without knowing why, I felt resentment.

The moment we walked into the visitor's area, sunny and carpeted and more pleasant than I remembered, the resentment returned. "Hello, Grandma! Merry Christmas!" Diana said cheerfully, hugging her. My father greeted her, too, more awkwardly, while I hung back. The room had other visitors, who seemed bored and anxious, caught here because they had to express holiday courtesy, and I felt depressed to be among them. But the people I resented were the residents, the old folks.

The dress Johanna had on, pretty and yellow, was awry on her shoulders. She had the loose look of a child. Her hands, when I took them, had a dull, papery feel. Her ravaged face seemed half asleep. ''Howard, Howard, how's my boy? Where've you been keeping yourself?'' she asked directly.

''Hi, Grandma. You look good,'' I said.

''I look old, but I feel good.'' She smiled and looked me in the eye, and her old inspirational self flooded back. She wheeled toward one of the tall floor-level windows and pointed a finger outdoors. ''Haven't seen snow like this since the Depression. Look at it.''

''We've been practically snowed in at the farm,'' Diana said.

''Wasn't anybody moving here either till the snowplows came through. Not that I minded, I'm not on the go anymore.'' She looked down at her ruined legs, then said, remembering Christmas, ''I don't want to be snowed in for Christmas, though.''

''We'll come and get you on the tractor if we have to,'' Diana said.

My father glanced down a hallway. ''I better go check with the head nurse to get it approved.'' He walked away.

Johanna pulled on my hands to have me kneel beside her. ''I'm glad to see you,'' she said, and added, almost confidentially, ''the next time I see you I'll be at home.''

Seeing her so lively made me feel better. ''What've you been doing with yourself?'' I asked.

''Oh, my. I keep busy.''

''Too bad they don't let you braid rugs here or plant a big garden like you used to.''

''Oh, I'm not up to that anymore.'' Her fingers fumbled in her lap.

''The important thing is you're doing something. Sewing, playing cards, it doesn't matter.''

She turned from me toward the window, clearly uncomfortable that I was pressing this point. But she had said she kept herself busy. ''I sit and read the Bible,'' she said. ''I pray someday soon I'll be with Pa.'' Her dress slouched farther off her shoulder. ''I'm glad you came.'' She was distracted. Her eyes dimmed. Her words came from nowhere. ''You be a good boy now. Mind what your folks say.'' I began to feel my resentment of her well up, not for admonishing me, but for being old. For so long she had seemed just a more fragile, graying version of the lithe woman who outfoxed us in our boyhood games and who tromped through swamps with us to pick huckleberries, who hoed in the fields with us, who sang songs, laughed, made us feel good. And then, while I was away, she had become old, and now she could be full of life one minute and emptied of its vibrations the next. I could feel her slipping from me, which for so long I could overlook, and I resented it.

....

The surface that the three chickens ran onto through the coop door was a sheet of snow briefly melted yesterday and refrozen in the night. The chickens slid and tipped, while my mother and I circled from opposite directions. Out in the cold the three chickens would not survive long. In the board game Trivial Pursuit, the question is asked, ''What is the dumbest domesticated animal?'' Answer ''Turkey,'' and you will advance in the game, but the chicken is a close runner-up. Jungle instincts have been bred out of chickens, and their own survival ability approaches zero, so expectant have they become on human service and security. At the top of a snowbank, my mother leaned toward a capture. ''Hold still, you bugger.'' The chicken pecked at her sleeve, striving for skin, and then gave up and sat in my mother's hands like an effigy borne in a procession. A second chicken rounded the icehouse and danced before me, left to right, right to left, thin legs frantic below fluffed-up feathers, completing in the snow an oval that suggested a race with herself. For no reason the chicken changed course and ran at me, achieving takeoff speed. She landed in my arms, and I returned her to the coop with the other. But the third chicken skidded into the open garden.

My mother and I stood and caught our breath. Diana was with my father in Midland, grocery shopping, and had I gone along my mother would have been chasing the chickens by herself. Years ago, when our house was crowded with three generations, she had wished for a little peace and quiet. Back then she had no true idea what solitude was. Now, suppose there was a major emergency with no one else around? Last August she had chopped the heads off two roosters, and, while pouring out a kettle of water to defeather them, had scalded one hand—a first-degree burn. She was alone, and the roosters had to be dressed out. ("I couldn't let them sit in the sun and spoil.") She had told herself the pain was somewhere else. This was one of the experiences, she said, that had reminded her she could invoke her own resources of strength.

My mother had grown up with three brothers and a sister in the third generation of the Buchhage farm family in Monitor Township, another of the outlying Germanic townships in western Bay County. In 1933, when she was fifteen, a terrible Depression year, she and her older sister were informed by their father they should make lives for themselves in Bay City. Separately, they became live-in domestics. ("At first, I was homesick. I suppose I cried myself to sleep a few times. And I had some narrow scrapes. I was out for a walk once, and two rough-looking fellows started to follow me. I hurried up, but they chased after me and came right up onto the porch. I pushed one of them down off the porch, and I got in the door and slammed it and caught the other one right in the face. I was shaking like you wouldn't believe.") That incident occurred when she was in the employ of Dr. and Mrs. William Kerr, in a stately Victorian at 730 Farragut that had the touches of a civilized imagination, marble and porcelain and velvet. Whatever jobs the Kerrs gave her, cooking, cleaning, gardening, nurse-assisting, she performed them, according to Dr. Kerr's thank-you notes, in the friendly, concentrated, getting-it-down-with-improvising manner that stayed with her. He soon was like a favorite uncle. The strangeness and trepidations disappeared. Thursday afternoons and Sundays were free personal time, and the feel of it, new to her, energized her to see the town. ("It was a beehive, a lot busier than now. You should have seen the fancy stores. Not Penney's and Kresge's and Mill End, but the really high-tone places with dresses from New York. I walked everywhere and saw everything.'') By the age of twenty, she was thoroughly competent in the city.

Now, in the garden, over which blew veils of snow, my mother and I tried to catch the third chicken, a poor strategy. In open terrain even a domestic chicken can be elusive. Disoriented, half hysterical, the chicken kept doubling back. It is all but impossible to guide a chicken. Flapping our arms might be misunderstood as a forthcoming attack; the chicken might die of fright. ''Scratch a chicken, and all you'll find is chicken'' is a farm saying, and it is the basis for a thousand dumb jokes about chickens, all of them true, and the basis for the several games of ''chicken'' that boys play. My brothers and I used to leap off the haymow ladder, from higher and higher rungs and greater angles of danger, backward, upside down, on our bellies, and, far crazier than that, I played country-road ''chicken'' with my teenage cars, first a Ford Fairlane, later a supercharged Pontiac GTO bought secondhand from the Tri-City Sunday drag races. This is the game of aiming your car dead-on at another, say half a mile away, and going at top gun toward a showdown until one or the other swerves. It is a game to horrify parents; as a father, I pray my kids never play it. Variations of it can be found across cultural lines and across the rural-urban line, but the idea behind it, holding your ground, is also pure Germanic. One time, when I was about five, a rooster tunneled into the pigyard, and I tried to recapture it only to come upon a Hampshire sow standing tall in the mud. Between the sow's molars an ear of corn snapped in two. I had seen her bite off the head of a king snake that she caught out looking for mice. The Hampshire lowered her head and kneed her front legs for a charge. There was a complete lull, a prelude to action, all eyes intently watching, even the rooster's, conveying from him an improbable process of mental reckoning. Then the rooster whizzed past me. Twice more the rooster escaped from a corner of the pigyard because the sow, baleful, making charge noises, overturning clods of mud, kept scaring me. Johanna was watching. "Go on out of there, boy. I'll catch that bird," she said, with a terrible, quiet click in her voice that sounded like the *tsk* of disappointment but which, I felt, was a larger shame. She climbed the fence and faced down the sow, imitating perfectly the angry squeal of a boar who finds someone in his wallow. Pigs get confused at a good imitation, and the Hampshire shambled off. Johanna's hand swept to the rooster and made a precise tackle, high on the drumstick.

My mother and I finally had the third chicken going the right direction. In the grape arbor, a clump of blackened, senile fruit drew the chicken into a series of hungry leaps. My mother and I moved in. The chicken jumped sideways into my arms. My mother judged differently and fell against the arbor's hang wires. Ligaments in her knee pulled away from bone. "You bugger, see what you've done." I could see my mother was in pain. But the rest of the day she acted, in front of my father, as if her leg was stiff, nothing worse.

....

In the living room there was a souvenir-stand painting of Texas. All Texas, in this milieu, was yellow sand, save for a long splintered tree, smooth as jade, made lifeless by the wind. I could hear that wind out the living-room window, a wind that could split a tree. It seemed to have blown a long way north.

"Another good day to be in by the fire," my mother said.

The wind had shrunk the big, roomy house. So much of every day we were confined to not just the house but to the wood-heated kitchen. The other rooms were too cold for sitting. I paced from one room to the next. The sense I had was of a ghost town. I made myself concentrate. These had always seemed lived-in, warm-looking rooms—flowered wallpaper and panels of pink linoleum in the dining room, lightened wood paneling in the living room and also in the first-floor bedroom where my parents slept, green-tinted wallpaper and linoleum in the downstairs sun room, a yellow-splashed bathroom, and upstairs, more wallpaper flowers and wood veneer in the four bedrooms. The couches, three of them, were wood and fabric. Every chair was comfortable. Colorful curtains hung by the windows. And on tabletops, from ledges, on the floor, above cabinets, was my mother's jungle of houseplants, her fantastic African violets, her shamrocks and spider plants and ivies and snake plants. Those nearest the windows she had moved toward the center of the house, away from the cold. I was shivering. I felt the cold through the panes of glass as if by osmotic decision. The cold had begun to feel too strong for me, stronger than the house and its people: the cold of irrevocable displacement.

"I'm going out to grind feed," my father said. "The chickens are running low."

"There's feed for three, four days," my mother countered.

"I don't want to be grinding feed on Christmas. Besides, I'm tired of being cooped up." He was halfway into his vest and coat.

"You going out too," my mother said to me. It was not a question. I put on bib overalls and a pair of my father's trousers that I had to cuff at my ankles and bunch at my waist—also, a coat, scarf, a wool cap down to my eyebrows, work gloves, galoshes.

In the tractor shed my father started the gray Ferguson, the smaller of his two tractors. If the ignition system failed, he could kick over the starter motor with a crank. He had paid $1,795 for it in 1953, and had paid $4,700 in 1966 for his second tractor, the barn-door-red Massey-Ferguson. Compact, manageable, modestly powered—the gray had a 35-horsepower engine, the red a 65-horsepower—the two tractors were of a type still profitably manufactured by Massey-Ferguson Inc. in Minneapolis, where the thinking is a generation behind other American farm-machine companies that sell principally eight-wheeled goliaths. "Farmers nowadays like those big tractors so much they must think they're going to drive them to heaven," my father says. Massey-Ferguson, bucking the trend, is a leader in worldwide tractor sales—number one, some years—but only because farmers in foreign countries prefer its small tractors.

On the metal Ferguson seat, padded with a blanket, my father shifted into reverse and backed up a low hill that bulged and crested at the barn. He parked. The front of the tractor, with its narrow gray engine hood inclined downhill, suggested the head of a mule. In the barn I slid open a double set of twelve-foot, sliding-track doors, and from the barn floor I uncoiled a pulley belt. The belt was rubber-and-canvas weave, eight inches wide, twenty-five feet long. I slipped one end over a pulley on the old hammer mill, the other end over a counterpart pulley on the tractor. My father fixed a jack at an angle behind the tractor and jumped the jack handle, forcing the idling tractor a few feet down the hill. On the two pulleys the belt tightened.

"Set the brake," he said. I jammed it down and notched it in place. I was about to shift the hydraulic lever and engage the pulleys. "Whoa, wait," my father said. "You forgot to get everything else ready."

Back in the barn we filled bushel baskets with shelled corn, wheat, oats and a Master Mix blend of bonemeal, salt, and other minerals. "I guess I'm paying twice over for the minerals now," my father said dryly. "I have to buy minerals to feed the chickens, but I have to pay to get rid of minerals from the water." He took delight in the irony. "I wonder if I gave the chickens water straight from the well whether they'd get their minimum daily dose of minerals."

"You should bottle the water and sell it."

"Fred Kohn's mineral water, huh?" He rocked on his heels and lifted a bushel of oats into the hammer mill.

The formula for his chicken feed was a half bushel of corn, a large scoop of wheat, a smaller scoop of the Master Mix, a bushel of oats, repeated over and over. The first repetition was ready in the hammer mill's metal trough. A hammer mill is a kind of elemental food processor. Its trough slants and narrows into a metal mouth beyond which, contained in a rounded steel head, are pulverizing sprockets like teeth. The sprockets reduce the grain and minerals to chicken feed. "Okay, start her up," my father said. The tractor pulley spun, and the motion carried along the belt to the hammer-mill pulley. It made a roar.

The ground-up feed flew from the main body of the hammer mill and flailed through alternating wind tunnels into cotton feedsacks. Small holes were worn in the sides of the tunnels, over which my father had wrapped strips of old feedsacks and baler twine, hardly airtight, and dusty clouds seeped through. The wind blew the dust in our faces. The wind was coming through the open double doors. I had closed them to within a foot of each other, leaving a gap for the pulley belt, but the gap might as well have been a canyon. The wind seemed to have eyes that found its way through the doors and in a wild rush found us. The wind became the prevailing noise, a howling, otherworldly noise. It was cold to the point of a coronary. This morning we had heard a radio announcer say, "If you *must* go outdoors today, please remember that overexertion in extreme cold greatly increases the risk of a heart attack." The overnight low temperature, with the windchill, had been fifty-seven below. In Detroit, a grandmother who fell and broke a leg had frozen to death on her front stoop. Her woolen coat had to be cut off with a tin snips, the wind had stiffened it so. Another Michigan woman, in jeans and a sweater, had ventured into her backyard to take out the garbage. The back door locked against her. The door and the wind muffled her cries to her two young children inside. Between her and the street was a tall wood-stake fence, gated, latched, and padlocked. She tried to climb the fence. By this time she had been in the cold several minutes, though perhaps less than twenty, in a medical examiner's opinion. She was found straddling the fence. She was twenty-nine.

At the hammer mill, inside my thick wool socks, my leather work shoes, my lined galoshes, I was losing the feeling in my toes. I might have mentioned this to my father, but with all the racket I would have had to holler. Finally I did. "How much longer?" My father tied off a feedsack and motioned to a pile of empty sacks. "We got these yet to do!" His reputation was that he did not leave a job unfinished. "When Fred takes on anything that's the last time you have to think about it," Reverend Westphal has said. For a time, my father was the Beaver Zion sexton. Preparing the graves, he used a shovel and a pickax, and even in winter no burials had to be postponed.

I kicked my boots together to stir up circulation, and we finished filling the empty sacks. Without a word, my father walked to the tractor, disengaged the pulleys, yanked off the belt and, on the open tractor seat, rode off. Sunlight and snow were both coming from the sky. I watched the tractor lose its shape, regain it, lose it again, giving his journey a movie aura of mystery and adventure. He returned on foot a few minutes later, while I was rolling up the pulley belt. Together we swept up the barn floor. Afterward, in the kitchen, we tracked over newspapers and warmed up against the stove. Snow, falling from his great eyebrows, sizzled on the cast-iron. The fire gave me back my toes. At the table, my mother had her roller pin out. She was enhancing with friction the consistency of tallowy cookie dough. Cookies from her oven were cooling on the gray surface of cereal boxes she had cut apart and laid flat. She did not remark on our entrance. My father said, "Came in to watch the cookies bake. Don't want to have to eat burned ones for supper." A smile crept over my mother's face.

The past hour, cold and miserable, had lifted my father's spirits. Forgetting about frostbite, I began to feel good myself. What is good about work, in the German Lutheran orthodoxy, is that you are allowed to savor and enjoy it and are forgiven, within limits, the sin of pride. The Lutheran work ethic is the fundamental religious underpinning of the Protestant work ethic. When the Roman Catholic priest Martin Luther broke with Catholicism in 1517, nailing ninety-five theses to a cathedral door, he rebelled also against the doctrine that work was "a scourge for the pride of the flesh." On the contrary, Luther wrote, work is "the base and key to life," and work itself, rather than "the fruits of labor," will set you free. This challenged centuries of Western teaching. Aristotle had said work was "a brutality of the mind."

To the Israelites it was "a painful drudgery" by sin condemned, or, to quote Ecclesiastes, "The labor of man does not satisfy the soul." For the early Christians, work filled in idleness, depriving the Devil of his time, but it had no inherent worth. The Catholic papacy made work obligatory for its monks but ruled that only prayer brought them to God. It was Luther who created a kind of Golden Age for workers, preaching that work elevated the worker and simultaneously served God. "With this, the great split between religious piety and worldly activity is resolved," C. Wright Mills wrote in 1950. "Profession becomes 'calling,' and work is valued as a religious path to salvation." I had studied Mills in college. Writing for the post-Depression, post-World War II generations, Mills tried to evolve a modern, socially correct attitude toward work. Work may be a lot of things, he decided—"a mere source of livelihood, or the most significant part of one's inner life"—*but* "neither love nor hatred of work is inherent in man, or inherent in any line of work, for work has no intrinsic meaning." My father had made a lifetime obsession of work. For me, Mills had diluted the obsession, although right now I could feel the Lutheranism in me.

....

Two chickens were dead when I went to the coop with my father in the afternoon. He dropped them by their feet onto a snowbank behind the icehouse. A dog across Carter Road howled and set off an answering chorus from, it seemed, every dog within earshot. One of the dogs, straying and inquisitive, might find the frozen chickens before spring, or else the carcasses would be thrown in a furrow ahead of a plow.

Snow had blown into the coop under the eaves and melted into the manure on the floor, squishing underfoot. We scattered a bedding of dry straw over it. "You can't win for losing," my father said. Yet he would not entertain the thought of a new, corrugated metal coop. It was my father's distinction that he had added to the old-style farm buildings but had not torn any down.

They had lasted out of care and luck—lightning had struck but never ignited—and because of their simplicity. The wallboards could have been sealed with a tributylin stain that will preserve wood for twenty years or more. A landlord of mine in San Francisco had put annual coats of tributylin on a wood balcony. After five years it was unsafe to walk on. It had rotted from the inside out; the preservative had sealed in the San Francisco fog. Without preservatives, the wood in my father's barn and sheds had dried hard as rock from the sun and wind. We used every one of his buildings. They were modest and slightly gawky and quite usable.

Two hours from Washington, D.C., an old German farm, its buildings intact, is open to public viewing. The barn is a holzsteiner, of wood (*holz*) and stone (*stein*). The historian James Westfall Thompson wrote, "The German was sure to build a large, fine barn before he built any dwelling house for his family except a rude log cabin." Thompson claimed that German barns described their builders: large, artless, thought-through, more complex than is apparent. One Sunday afternoon in summer Diana and I had visited the old farm. The parking lot was lined with cars. We paid an admission fee and entered through a turnstile. A farmer stood near the entrance to answer the rudimentary questions of city people. He was a young, lanky man with thick woolly curls, wearing a blue plaid shirt, so self-possessed this farmer, so faithfully dressed, so humble and forthright in his speech, it took me a while to understand he was an impersonator, employed only as a tourist's aide. In Washington, on weekdays, he had a job in an antique-frame shop.

"Your dad must've been a farmer, or your grandpa?" I said.

"My grandpa was, but he died when I was real young. I can't say that I remember him."

Diana and I wandered around the farm, which was fenced in to keep out anyone who did not pay the fee. I went into the barn. It had an old look on the outside but the floors and rafters inside were an unweathered yellow. The barn was not old; it had been made up, like the farmer, to look the part. The walls outside had a chemical smell. An antiquing chemical had hastened the graying and cut in rough lines. The tourist's aide was at the entrance, explaining that Germans were a decided, steadfast people who left a place as the last resort. "They came to this country because of great duress in their own country. Their land was taken from them. They were forced into cities. Their choices were limited—be wage earners or soldiers or the wives of soldiers, or find new homes somewhere else." When he was finished, I asked where his interest in farming came from, if not from his family line. "I've read a lot," he said. He mentioned books by Robert West Howard, Wendell Berry, Mark Kramer. This started him off on a long lecture, and when he paused, I suggested that his true calling might have been as a professor. "No," he said. "If I could afford it, I'd actually like to farm. Maybe it is in the genes."

....

At dawn the temperature was forty degrees above zero. By ten o'clock it was fifty, a heat wave. The ditches ran with a month of snow. "Looks like a good day to split a few blocks," my father said.

Cra-pop! Cra-aa-pop! Blocks of wood fell open in half-sections and quarter-sections. Before long I had a sweat. The low-pressure zone that was warming the air was also making a mess of the ground. The winter turned to mud. The footing became unpredictable.

"Take a break," my father said. I leaned my ax against the barn and stood a moment on round, smooth stones over which running snow ceaselessly played: a standing area for cows that Heinrich and his sons had laid. The stones were collected a few at a time, tossed from the fields into piles under shade trees, carried from the creek bed in the dry season, brought to the barn on a flat wooden sled, the men singing to the horses that pulled it. Heinrich, skilled with bricks and cobblestones, showed his sons how to set the fieldstones in earth, forming a square next to the barn like a courtyard before a palace, but one now without occupants. Inside, the cow stanchions now hung every which way. They were a loose-swinging plaything for the grandkids, who could run up and down the manger staging war games with plastic swords and guns.

While my father and I rested, my brother Roy drove into the driveway, and his daughter, Renee, too young to remember the cows, jumped out and ran over. ''Hi, Grandpa? What're you doing?'' she shouted. Her boots splashed the heat-wave puddles over the stones.

My father was sitting on the flat edge of an old stone-and-cement water trough. ''I'm supervising!'' he said in a mocking voice.

''Hey, Merry Christmas!'' Roy yelled, walking up behind Renee. He gazed out on the liquid landscape. ''Or should I say Happy Easter?''

He stood next to the old trough. ''Where does this crazy weather come from? Every year it gets crazier. There's no rhyme or reason to it.''

''Write the CIA. Tell them to go experiment with Russia's weather instead of ours,'' my father said. The Michigan weather, inconsistent, mutable, unsettling, is never anything less than crucial to a farmer, but on a day like today my father could joke about it. Smiling at Renee, he indulged her as she hopped and skipped over the slippery stones. ''Who's going to pick you up when you fall on your face?'' he said.

Roy sat down. ''Don't let us get in your way, we just came over to supervise,'' he said, cribbing my father's line.

Of our generation, Roy looks the most like the historic Kohns, like Heinrich, like Johann: the full, confident build, deep chest, big shoulders. big, sinewy hands, red, healthy color, skin smooth and tight. He lives with his wife, Lorie, a short, energetic, pretty woman, the only serious girlfriend he ever had, in a fine new house on the back forty. Roy alone had always kept close to home, living nowhere but on the Kohn farm, although in his early twenties he nearly went off farther than any of us—to Alaska, with the construction gypsies that put up the Trans-Alaskan pipeline. By inclination he is a fine-skill carpenter, and, in his basement workshop, he had a sideline business of handcrafted breadboxes, quilting frames, and other wooden items that he sold to a waiting list of customers. His steady job was at the construction site of the Midland nuclear plant, where he operated heavy machinery.

"Daddy's got today off, and all next week. Isn't that super?" Renee bounced gleefully back and forth between the barn and my father. He had succumbed by now, quietly, to being upstaged.

"I could use a month's vacation," Roy said, "with all the overtime I've put in."

"I'm surprised they didn't tell everybody to stay home permanently," my father said. This was rhetorical ridicule, but also close to the truth. The nuclear plant was about $2 billion and five years beyond its original guidelines with no end in sight. Imperfectly conceived in the early 1970s—the quicksand nature of the sandy soil not taken adequately into account—the structure began to sink before it was half completed. A flotation ring of cement, attached underneath to redistribute the weight, failed to stabilize it, and engineers could not swear to its future position. My father considered it a white elephant. "Who's going to pay for that two billion? You can bet it's not going to be the jerks who made the mistakes. The way this country works, the bigger the mistakes you make the easier it is to get away with them."

Roy, with a well-developed sense of the Lutheran work ethic, had nearly walked off the job his first week at the nuclear plant. He was being paid to stand around, and one supervisor even told him to while away his time at the bottom of an empty water tank. Finally he was given a wheelbarrow and told to gather up broken chunks of cement, which satisfied him. Now he had seniority and considerable responsibility.

Sandra, who also worked at the nuclear plant, had even more. She was a senior bookkeeper in the payroll division, and had been one of the first people hired after groundbreaking, nine years ago.

No one could feel good about the new rumors that construction might come to an immediate halt—a *Midland Daily News* headline this week had wondered: IS ABANDONMENT ONE OPTION? Without Sandra's job, the rest of Mike's education was up in the air; all their plans were. Just yesterday Sandra's supervisor, proud of her, had approached her with a special offer, a transfer to another nuclear project in Ohio. "It's nice to be asked," she had said last night. "They'd pay my moving expenses: it's real preferential treatment. But I'd have to go almost right away. Once it hits the fan here and they start laying off, everybody will want transfers. Some girls already have applied."

Sandra often rode to and from the nuclear plant with Roy, and my father asked him what he guessed Sandra might do. "I imagine she'll talk it over with Mike," Roy said diplomatically.

"Wouldn't surprise me if she takes the transfer," my father grunted.

He pointed with his ax to several blocks of wood that he had separated out, pieces of upper tree trunk from its main intersections. "They're too knotty for me to split, so I saved them for you," he said to Roy. "Think you can split them?"

Roy is big and strong, the only one of us who on first glance is obviously bigger and stronger than my father. Even so, I had not known my father to be this open about it. Roy took my ax from against the barn, and, with a slightly nervous aggression, he swung into a knotty block.

"No, no," my father said—a prolonged command. "I told you, they're too knotty to split with an ax. Take them with you and use your hydraulic splitter." Roy, at his house, had a motor-powered device that drove a wedge on a four-foot steel arm, greatly accelerating the speed at which he divided his woodpile into fractions for his two fireplaces and a wood furnace. "Okay," Roy said, "I'll split them for you and bring back the wood."

"Nah, you keep it," my father said. He began to roll one of the blocks toward Roy's pickup. "I've got plenty."

My father's mood, a wave rising and breaking, caught us

flat-footed, uncertain. Roy and I stood a moment in the hot winter sun, which was something else out of kilter, and then we loaded up the blocks. Bending, lifting, keeping the weight of the blocks low, heave-hi-ho-ing we piled them onto the pickup bed. I put my arms around a fat, bumpy block. The hoisting and straining felt good. It had the feel of health. Deskbound muscles were loosening up. I was beginning to believe I could lift any size block. But the next one was a two-man block, and it staggered me, knocked me off balance. Roy steadied me. "Give me one end," he said. We rocked it between us, gathering momentum, and swung it with a foot to spare over the tailgate. The quarter-moon scar below Roy's left eye was turning a brighter red. The scar had come from a fall off his bike into a barbed-wire fence when he was nine and I was sixteen, on a day my father had left me in charge of the farm. Roy found humor in the incident now, although he had nearly lost the eye.

The pickup was filled to its limit. Renee disappeared into the house to find Diana. Roy, taking a third ax from the toolshed, said, with a cockeyed grin, "Since I'm on vacation, I might as well have some fun." In a minute, the three of us were swinging axes to a steady beat. The remaining blocks flew apart.

A year ago Roy was on an injury-reserve status at the nuclear plant, not sure if he would ever again be able to handle physical labor because of a mystery disease that began with flu-like symptoms and deteriorated rapidly. In the middle of Lorie's sleep one night, Roy woke her, mumbling, and, draping his weight over her shoulders, she piggybacked him into the car and drove full tilt to the Midland Hospital. His lower legs were ballooned up; he had blood in his digestive tract; an unknown poison (bacteria? virus? chemical?) was ripping through him. He was on the critical list with the outcome in doubt for several days. The saving power was Roy's youth and reserve of strength. His attending doctors were at a loss for a diagnosis and a treatment. Tests were to no avail. For several months afterward he could not get rid of an arthritic pain in his knee and ankle joints. Some days he had to stay in bed; other days he hauled himself to the nuclear plant, managing on six aspirin every few hours. He had the force of a weakly wound toy. At last the pain began to subside, either of its own accord or, as Roy thought, due to the vitamin remedies of a neighbor who had studied "natural medicine" books.

"Are you completely recovered?" I asked.

"I probably have two or three bad days a month where I run through a lot of aspirin. Cold seems to make it worse. Today—today is a good day!"

We were down to the last few blocks from a pile that yesterday looked like it would take all winter to chop up. "I'll call Ronald tonight," my father said to Roy. "He's off for a few days too, and with this weather we might as well buzz wood at your place tomorrow."

"Sounds good to me."

My father put aside his ax and, bending, picked up the split sections and laid them in parallel rows. At the ends of the rows he crisscrossed the firewood for greater stability. My mother, her hair loose, coat unbuttoned, boots flopping, came from the house across snow that had grass poking through. "It's warmer out here than it is inside," she said. She began handing wood to my father for his methodical rows. *And the Lord God said, It is not good that the man should be alone; I will make him an help meet for him.* In my catechism class, teaching us the Lutheran concept of the Beginning, Reverend Reimann quoted the original Hebrew: *azer,* for "help meet," a unisex word: to give assistance, as someone who works side by side. Eve was brought forth out of Adam not to be his lover, as the liberal theologians had it, but to be another worker. Work had preceded love. Work existed in Eden before the Fall, when the Lord God's creation was in perfect order. Work was not a consequence of sin. It was Aristotle who had joined them and made work unholy.

For another hour we chopped and piled wood, uttering not a word. The work swallowed us up. Language was unnecessary, an intrusion. When we were done, Roy said he had to leave. The miraculous shining December day was departing, too.

....

The next day was back in its season, and Roy's knees and ankles hurt. But by 9 A.M. Ronald, Mike, my father and I had gathered at Roy's house to buzz-cut a large, unwieldy pile of logs. My father selected a flat area near the pile, and Ronald and Roy lifted the buzz saw out of my father's truck and set it down. I backed the gray Ferguson into general alignment with the saw's flywheel pulley. Mike fitted it with a short canvas belt. Shoving, backing up, yelling out, my father worked with Ronald and Roy to maneuver the perpendicular of the belt and the saw into a crisp right angle. "A little to the west. No, too far. Go slow with it. Hold it, hold it! Okay, get the spikes!" Four railroad spikes would secure the saw to the ground. Roy leaped in and out of the truck to get them. Landing on his feet after a momentary free fall, he flinched. "Think I'll take a few aspirin," he said, and ducked into the house.

"You know he's not the only one around with this disease or whatever it is," Ronald said. He named three other cases that had come randomly to his attention, a twelve-year-old boy, a middle-aged woman, an older man. Their symptoms were the same—the punishing, life-threatening flu and the lingering arthritis—and they lived in proximity to the oil drilling and drank well water.

"It could be the water, not that you'd ever be able to prove it," I said. I knew this much, from phone calls in the past few days to environmental researchers in Washington: oil and gas drilling in seventeen states, Michigan included, was blamed by the Environmental Protection Agency (EPA) for contaminating rural drinking wells with toxic metals such as mercury, lead, and chromium, and also benzene, formaldehyde, and stronium; in Ohio, state officials had received more complaints traceable to oil and gas drilling than to any other source of groundwater pollution; some of the pollutants were chemicals called "drilling muds'' that made drilling easier, but most occur naturally in underground formations and were surfacing with the oil and gas; when not properly disposed of, they were leaking into wells. But cause-and-effect on an individual level was difficult to establish. Still, one of the cases Ronald mentioned was intriguing. He was Ed Zimmerman, hospitalized for several weeks, whose son, Cliff, was Ronald's partner in men's horseshoe doubles. Cliff, who farmed the old Dumont place directly across Carter Road from the Kohn farm, had become known as ''The Millionaire.'' Three wells had been drilled on his property, and all three were producing. ''This thing just hit his dad out of nowhere,'' Ronald said.

''Well, we've got plenty of pollution around,'' my father said.

Nearly every day this year there had been a story in the *Midland Daily News* about the dioxin TCDD and thirty-nine other chemicals that the EPA said was going from Dow Chemical's discharge pipes into the Tittabawassee River. In the year-end roundup of big news stories that the Associated Press had just compiled, the story of the poisoned river was up near the top of the list. Everyone had been warned by the EPA not to drink the water or eat fish from the river, but not everyone was conscientious.

''Okay, okay, let's go,'' Roy shouted, returning from his medicine cabinet. ''Why are all you old ladies standing around?''

The pulleys spun. The belt pulled straight. Ronald and I laid a log, butt end first, on the saw's cutting tray. Roy leaned his weight against the tray, which swings back and forth on a V-hinge, and the log hit the gap-toothed buzz saw. A two-foot block, ready for splitting, fell off the tray into my father's hands. He flung it to the ground ten feet away, the start of a new chopping pile. ''You want me to do that?'' Mike asked, but my father gestured him toward the log pile. Mike and Ronald and I alternated with the logs. We tugged them out carefully lest we set off a rolling avalanche. On some the bark was soft and corrupted, unreliable in our hands, but, from memory, I seemed to know exactly how to handle them. I moved with the rest of the guys, fitting in, all of us moving as if obeying one marching order.

From early on, we had worked with wood. We knew how to cut down a tree, and also how to build things of wood, milk stools, a toolshed, a house. In 1957, when Dale was born and the farmhouse finally reached overcrowding, we built the retirement house for Johann and Johanna on the back forty. Knowledge of wood was part of the Saginaw Valley, part of what was handed down. Johanna's grandfather, John Stephen Walter, had been the chief carpenter in the construction of the Beaver Zion church and had owned and operated a sawmill in Beaver. In 1904, when Johann was fifteen and already out of school six years, he took a job at the Walter sawmill, and he posed for a picture with his brother and two friends from the mill crew. The camera lit up the boy-men with their strong, oily faces, their devil-may-care eyes. They were in sweated shirts and suspendered pants. Beer kegs were in the picture. Johann is holding a beer stein. He dominates the picture with his serious swagger, a grown-up at fifteen. He was paid $3.00 a day at the sawmill, about twice Heinrich's wage in the 1870s. Johann was working to buy, in his turn, the Kohn farm. It was a reckless thing to do in 1904. Not to put your savings into farming—which, however hard on the muscles, however connected to the threat of beetles and fungi and freak weather, however full of tough breaks, was still very much a good living—no, not to buy a farm, but to work at a sawmill. Johann got his job because another young man had a thumb sawed off. At the turn of the century every sawmill worker could expect something to go wrong at any moment—the suddenness of trouble that is peculiar to a declining industry. Things, when they are going bad, can get worse in seconds. Machinery was antiquated and dangerously worn. Saw blades cracked and flew apart. Gears snapped; pinions split, collapsing the saw trays. There were fires. One worker, perhaps horsing around, fell into a dune of sawdust that was burning hot inside. The heat reached air and roared fifty feet up the dune in flames. He could not climb out of the encircling wall. There was an air of doom about the sawmills and a rubbled volcanic look, a sense of impending eruption. In 1909, at J. C. Ittner's sawmill, the biggest of the Beaver mills, a steam boiler exploded, and the mill burned. Explosions and fires one after another destroyed the sawmills all along the Saginaw and the Tittabawassee, reducing the 112 mills of 1882 to a handful of survivors by 1910, the year the Walter mill was wisely shut down, and four years before Johann, a prospering full-time farmer by then, married the owner's granddaughter.

....

At noon the Dow Chemical factory whistle hooted across the countryside. In the din of the buzz saw, we didn't hear it. At five minutes after noon my father looked at the sky and said it was time for dinner. He drove to the farmhouse and brought back Diana and my mother with a pot of chicken noodle soup, which we ate alongside the full meal Lorie set out. The five men ate in gulps, although, by dessert, we were more or less mannerly. My father became talkative. There was a smirch of tree grime on his forehead, and his hair lay fitfully, giving him a jaunty, boyish air. He was in what Diana calls his "news reporter's mode." He told us of a recent meeting about the million-dollar centennial project at Beaver Zion that had grown heated. "I got up and told them, 'If we're going to raise a million dollars, we need some incentives. When I was in Texas last year, I went to a church where they put cushions on the pews, and, as soon as they did, their offerings went up by thirty percent. So maybe we should put cushions on our pews.' Well, some guys laughed, but some got even madder because they thought I was making fun of them."

Granted it was humor with a little dig, but humor nonetheless, from someone who, I once thought, did not have any sense of it, who seemed to have on his mind one unleavened, unrelenting subject alone: farm work. A half hour like this would have been dawdling. Always there was work to do, and always it came first. Before breakfast, we had to feed and water the livestock, the cows, the heifers, the bull, the pigs, the chickens. We had to squat on milk stools, wipe clean each cow's teats, squeeze chromed pails between our knees, swish away, dodging the cow's tail, which could slap you silly; and in winter we had to shovel out the manure trenches. It was boring to start every morning this way, and boring to hear at breakfast your quota of work for the next five hours and to hear at dinner your quota for the next five and to know that after supper you would end the day as you started it. It is not easy to be in such a family, nor, I suppose, is it always easy to head up such a family. You must have a formidable attention span. From all I had been told, Fred Kohn at age fifteen was exactly like Johann Kohn at fifteen. Was I at fifteen so much different, or was it possible that I had not seen my father fully, that I had not realized he, like everyone, had other faces? I didn't know. He wasn't clear to me, even now.

"Boy, that was good exercise this morning. Took all the kinks out of my back," my father said. He pushed away his dessert plate and stretched.

Outside, back at the buzz saw, he took up the same station. Some of the blocks weighed close to the limits of his strength, but he kept at it. Grayish clouds swept across the treetops like a line of faded laundry being blown dry. From time to time, though not with any heat, the sun shone through. About two o'clock it began to snow. Soon there were little snowbanks and snowdrifts on our caps and coats. White flakes slid in view-obstructing streaks down my father's glasses. Roy paused at the saw, but my father said, "Hate to quit when all we've got is an hour or so left." He wiped his glasses, and we shook ourselves, imitating animals, and turned up our collars. By four o'clock we were finished with the hoisting and sawing and flinging.

Mike left in a hurry. He had another project—Scott's wooden soapbox racer—waiting at home.

"I wonder where he and Sandra will end up," I said.

My father was banging on the railroad spikes to free them from the ground. He looked up. "Maybe they'll go see the world like you and Diana." He said this with more humor than I would have guessed.

....

Before supper there was a phone call from the Colonial Rest Home. Johanna had been taken by ambulance to a hospital as a precaution against fluid in her lungs that was advancing toward pneumonia. Supper was eaten in virtual silence. The first thing in the morning my father asked Reverend Westphal to pray for her at next Sunday's service.

FOUR

In the age of psuedo-holidays, Christmas was always a real holiday at the Kohn farm. The day itself was a pandemonious stirring of memories, and it had the kind of rituals that forced you to recall and compare each new Christmas to all those in the past. And yet, after we were grown, Christmas only rarely brought us together. The first one I missed was in 1970, when, as a Detroit newspaper reporter, I volunteered for a day on the police beat, another break in the tradition, another act of renunciation, even though I said I was doing it simply for the overtime. A few years later Harvey took a manager's position with Miller Beer in Fort Worth, and, with his workaholic habits, he and his wife, Carolyn, also became no-shows at Christmas. Dale was next, after he met Katherine in a cafeteria at the Johnson Space Center and they decided to settle in the Houston area, he in computers, she as an accountant. Only once in the past decade had we all happened to be home at the same time for Christmas. This Christmas—finally, a family reunion!—had been planned for three years. We had begun to talk among ourselves that we had better make an effort, ''or else it'll be somebody's funeral before we're all together.'' As it was, one thing or another had screwed up the plans until this year.

''Is there room for everyone upstairs?'' Diana asked.

''Harvey and his gang will sleep at Sandra's,'' I said.

''Sandra's place will be warmer for the kids,'' my mother said, adding her footnote. Her eyes pressed shut for a moment. All day, folding and unfolding her hands, she had muttered apprehensive reminders to herself that this big family event, so close now, might yet by some accident elude her.

Minutes ago my father had left for Tri-City Airport to pick up Harvey and Carolyn and their two children, Cindy and Alan. But Dale and Katherine were not due from Houston until later tonight, and another snowstorm was on the way. My mother could imagine them stranded all night in Chicago or St. Louis. Last night, when Sandra picked up her weekly box of eggs for her customers at the nuclear plant, there had been a brief moment of misunderstanding in which my mother thought Sandra was leaving today for an interview in Ohio about her job transfer, and my mother saw Sandra snowbound on Christmas Day in an Ohio motel. ''No, I'm not going to Ohio unless I decide I want the transfer,'' Sandra had explained, ''and I'm not sure I want it.''

(To me, the logic of the transfer was inescapable. In Ohio, Sandra would be guaranteed a paycheck while Mike finished school. Here there were no guarantees; her job was week to week. Not to mention that Ohio would be for her an expeditious way out of here.)

The snow held off. The planes carrying the Kohns from Texas landed safely, and the next morning I brought my daughter Liz to the farm from her mother's house. Once the family was gathered, I had to fight hard against my own premonition. Digging in me was the thought that Johanna would die, or my father would, somehow obliging us, taking macabre advantage of the fact that we were already here and wouldn't have to make a special trip for a funeral.

At the hospital, where Johanna was on the critical list, the scene was one of hovering concern. The technicalities of diagnosis and treatment were shushed back and forth, with death all the while in the air. ''She'll be okay. Hard to kill a Kohn,'' my father reassured us. He looked at me. It happened to be one of my moments of extreme dread, and I was unfathomably grateful.

''Remember,'' he said, ''the doctors thought your grandpa would never give up the ghost.''

Johann had lived through two heart attacks. The first was when he was seventy-seven. The doctors told my father to select a casket, and he called me to be a pallbearer. I hitched a ride home from the University of Michigan and found Johanna baking pies—for the funeral supper, I assumed. ''No, for Pa, when he gets out of the hospital,'' she said. The pies had to wait eight days. I was at their house the day he was to be released. She cut the rhubarb custard. Her hands shook, and she was breathless. Johann, bracing himself on a cane, came into view up the kitchen steps. He had deteriorated in appearance, his skin drooping from his cheeks, his suit a shabby fit from the weight he had lost at the hospital. His gray-white hair was parted over his left ear and the gay curve in front had been drawn flat with a wide-toothed comb. But he had the smell of Mail Pouch and liniment, his smell. He took my grandmother's hand and sat down heavily. ''Look at you, Ma. Just look at you.'' I saw that I should go outside and leave them alone. Later we ate the pie.

....

On Christmas Day, in with the readings from Luke and the ''Alleleuia'' hymns, Reverend Westphal offered a prayer for Johanna, and he closed with a prayer for all of us. ''May the Lord lift up his countenance and grant you peace.'' After that, he broke a little from the formality to welcome those of us who had come home to be with our families. In the chancel, two white pines stood blinking. They smelled of the woods but, in fact, had been cut from Loretta Casey's yard. At thirty feet tall, they were short and didn't really fill the soaringly airy chancel. To find Christmas trees that fit the church these days, someone would have had to drive out of the valley and deep into northern Michigan.

The organist played ''Hark the Herald Angels Sing'' as we

filed out, and at the carved doors Reverend Westphal reached out to shake spiritual hands, wishing each parishioner a better year ahead, saying it loudly, as though addressing the full pews.

At my turn he said, more subdued, "I'm sorry your grandma will miss Christmas with you." Then he brightened. "All things considered, you should have a pretty good celebration, though."

"Yes," I said. "I think so."

Outside, on the cement walkway, a woman waved and moved toward me. "Hi, stranger," she said. This was Jean Pawley, who had been Jean Ittner when I had a wild crush on her in the eighth grade. With a joyful air she grabbed both my hands, and, after I finished introducing Diana, she grabbed both of Diana's. Jean was the mother of two teenagers and did not try to hide it, but she had retained a lot of the charm and liveliness of the eighth grade. "It's wonderful to see you," she said. "How long will you be around? Long enough to stop by the elevator, I hope."

"Absolutely," I said. "Got any calendars left?"

"I'll find one for you." Ittner's elevator was one of the places that still gave away promotional calendars. Every year my father hung his from a nail in the kitchen. It had twelve fairy-tale photographs of America above a printed advertisement, "ITTNER BEAN & GRAIN, INC. Beans—Grain—Chemicals—Fertilizer—Seed, 312 Park, Auburn MI 48611. The owner was Oscar Ittner, Jean's father. Oscar and Jean had one of those special bonds. They bragged about each other, and when you saw them together, their faces merged: bright and round and cherubic like the sunshine faces children draw. Their differences—he is reserved and precedes his opinions with careful, preoccupied silences; she is demonstrative, ebullient, as if she has a hot line to everlasting good news—did not cancel out their closeness. But in 1967, after Jean graduated from Bay City Central, a series of turbulent, wholesale changes almost did. She married Tom Pawley, a friend of mine, who enlisted in the Army and served his time in Germany, from where he sent for her. Oscar wanted Jean to play it safe and fly all the way, but she rode on buses and trains so she could sample a little of the cities she was traveling through. She had not been farther than Detroit before, and, I believe, Oscar was not certain of her return. Then, while Jean was in Germany, the life she had known as an Ittner turned around. A letter from her father one day informed her that he had given up the Ittner farm, one of Beaver's originals, and, with no more than a passing knowledge of the Chicago Board of Trade, had gone into the elevator business. I saw Jean shortly after her return from Germany. She smiled and chattered deceptively, but I recognized beneath the smiles a look people get when they need to cry. And she admitted, "That's what I did when I got that letter, just cried and cried. I was so stunned. Usually, overseas, we were so happy to get letters, but with that one I went from a total high to a total low. I just cried. Dad's going to leave farming? How can he do that?" The next time I saw her, a year or so later, Jean was rejoicing in a state of chagrin. She had become her father's bookkeeper and right-hand assistant, and her husband, Tom, was hauling grain and beans for the elevator. Jean and Tom had broken ground for a new split-level house a stone's throw from the Ittner homestead. "I've eaten my words," she said. "The elevator business is ideal for me. There's always someone to say hello to. I've realized that I'm probably not cut out to be a farmer's wife. I like people too much." At Ittner's elevator, fifteen years later, she said hello every morning not only to her father and husband but her brother and one of her sisters, all employed there. Today she was alone at services, alone with her two children. Tom was home sick with a cold; her brothers and sisters were with their in-laws. "How's Oscar?" I asked, and her reply cost her an inordinate amount of her smile. "The truth is, his heart is bad," she said. "He get these pains which scare Mom, even though he tries to hide them."

"Same with my dad," I said, and tried to be cheering. "We probably shouldn't worry. Chances are they'll outlive all of us who sit on our butts for a living."

"True, true. Your dad, I think, will farm forever. He sure doesn't act like he's ready for a rocking chair." Her face with its nimbus of brown curls was alight again. "I hear he won't let the oil drillers near his place.'' Jean's mother, Luella Ittner, was a regular with my mother at the church quilting bees.

''Yes, he's making trouble, as usual.'' I felt a funny twinge of pride.

''Does he have someone to take Don's place?''

''Gerry Radke.''

''Too bad.'' Jean teased. ''I thought maybe we'd see you out there next spring.''

''That'd be the day,'' I said. I guess I wanted her to think I was still the same guy who had gone off to jet and jazz through life. Anyway it was how I felt at the moment.

''Well, before you leave for parts unknown again, come and see me. Business is slow. We'll have plenty of time to catch up,'' she said.

....

''Merry Christmas!'' At the front door there was shouting and the noise of Sandra and Ronald and Roy and Harvey and Dale and everyone else coming in, not bothering with the doorbell. There were coats to be taken off, and boots, and hugs and kisses to be exchanged. The push of people forced us out of the kitchen and into the other rooms. This big house of things accumulated over the years and lovingly maintained—three brown-tone couches, their wooden arm rests darkened with use and polish; one of Sandra's cornhusk wreaths tied with a red ribbon and displayed on a door; ceramic figurines of a farm woman gathering eggs in an apron and her man in patched jeans and straw hat; a varnished gag picture of two preschool coveralled boys in which one, thumbs hooked in his pockets, says to the other, ''You been farming long?''—this big house barely had floor space for all of us. In the middle of the dining room sat the long oak table on its eight cross-connected legs. My father had extended it to its full length, laying in seven auxilliary leaves with their metal probes and matching holes. The table top, a lustrous gold from his refinishing, was spread with white linen, an old wedding present but still showy, no stains. The centerpiece was a display of pine cones and ceramic bells sprayed with sparkling dust.

"Count the places," my mother said. "Are enough places set?"

"Twenty-three. We counted them twice."

The fourteen adults were to sit at the oak table, along with Liz, almost an adult. The other eight grandchildren had places at two card tables spread with oilcloths and squeezed into corners. The women, my mother and Sandra and the sisters-in-law, carried in the food—dishes of corn and peas and carrots frozen from the garden last summer, and two roasted roosters also out of the freezer, and a beef roast, mashed potatoes, rolls, sliced bread, dill pickles, sweet pickles, a taco salad, a nine-flavor garden salad, two cottage cheese in Jell-O salads and a fruit salad, with dessert to come.

"Left the cranberry sauce in the kitchen," my mother said.

"I'll get it," Roy volunteered.

"Let's pray first," my father said. In unison, half under our breaths in a baroque pell-mell tempo, we did. Diagonal shafts of winter light threw sun into the room, glittering and collecting on our plates.

"You know what we had my first Christmas away from home? Canned beef from Argentina," my father said. "The beef tasted just like the can. You could hardly eat it with ketchup, and we didn't have any ketchup." He smiled. "This was in Tunis in 1942. Boy, it was hot, no snow, no wind. We'd shipped in from Britain. Had even worse food on ship. Whitefish in canned milk for breakfast, and canned crackers." I had not heard this before. The stories my father told when we were young and wanted goose bumps were about black bears in Michigan's woods, not about the faraway war in lands that he, as I thought, had forgotten. England, Algeria, Italy, and Sicily were where he had served. "In Algeria," he said, "we outfoxed Rommel. The Germans had us outnumbered by three to one, four to one, but the American commander kept us on the go so it looked like there were more of us." Moving in circles to imply numbers they didn't have, the Americans marched on red clay roads, brittle as pottery. Their thoughts were in their stomachs; the canned crackers and Christmas beef had not gone down well. Off duty, they resorted to other deceits of wartime. One GI stood on the front bumper of an Army Jeep, traveling at road speed, behind a French supply truck. The Jeep motored closer. The GI opened the rear door of the truck and snatched a bag of flour. A pastured cow was found and shot and the owner reimbursed after the fact. My father, trained by the Army to cook, baked biscuits; and, already knowing how to butcher, he cut up steaks. "And we bought oranges from kids on the street. Big oranges with pink insides, sweet as honey."

My mother cleared the table, and desserts as quickly filled it—pumpkin pie, cherry pie, apple pie, vanilla and strawberry ice cream, blueberry cheesecake, three sideboards of cookies. We should have prayed twice. Gluttony was inescapable.

We reduced the feast to a few sugar cookies, and the grandchildren sallied forth to positions near the lighted tree, a short needled pine with expansive branches that for all its expansiveness could not shelter half the gay-looking presents. "Okay, settle down," my father shouted. The devil was in some of the kids.

My father began dressing for the chicken coop, and Diana tugged at me, "Why don't you do the chores and let him stay indoors? It's Christmas."

"He wouldn't let me."

"Even on Christmas? Go ask him!"

"Believe me," I said. "Helping him is one thing, but doing his chores for him? Not a chance! Not even on Christmas." I was suddenly restless, though. "Look, I'll help with the dishes."

"You?" Diana laughed. "Hey, Sandra, come get a picture of this."

It was a family joke whose time had come: the men washing the dishes. Margo handed a dish towel to Ronald, who had to be led by the hand to the kitchen. But when Roy fell into line without protest, so did Harvey and Dale and Mike. "You boys don't have to,'' my mother said politely, and then, as Roy took the kettle of scalding water from her hand, ''You better let me do this. These are my wedding dishes. You got to be careful.'' Her concern rose less from the endangered dishes, I thought, than from the crowd of men displacing her from her kitchen. ''All right, all right,'' she said, seeing it was no use, joining in the joke, and with a slapping of her bunion-distended slippers on the linoleum, she went into the living room. Roy poured water from the kettle into the dishpan, raising suds, and stuck his hands in. ''Dang, it's hot.'' We laughed, and Sandra, laughing harder than us, snapped away with her camera.

''What'd you think? Will Zion raise a million dollars for the centennial project?'' Harvey asked Roy.

''No problem. All we need is for someone to win the lottery.''

Dale, who lived south of Houston, said he had joined a new, oil-wealthy congregation, exploding with members, that was about to divide in two. The largest half would move into a new church. The property alone might cost two or three million dollars.

''You ought to send a little extra cash Zion's way,'' I said.

''Hey, we've got our own oil wells,'' Roy said. His voice hit a high, light note, but among Zion's older members the question of paying one's own way was a serious, prideful matter. The centennial remodeling project had run up against it. The project (''so Zion doesn't fall totally behind the times'') had started with a budget triple the amount of money normally tithed at Zion for a full year of operations. The centennial committee, on which Roy served, had been caught between two philosophies: one, in favor of deficit spending, put forward by men who tended to be younger, educated and employed in Midland, typically at Dow Chemical; the other, by the patriarchs, who opposed debts and staunchly so. From Roy and several more committee members, a compromise was now in the offing. All the parishioners would be asked to make a pledge, committing them to pay so many dollars over so many years toward the project, and on that basis a budget would be finalized. The patriarchs had their reservations, though.

"If everybody in the congregation stayed put and didn't move, okay, that might work," my father had said. "But you can't count on that anymore." Newcomers to the congregation presented the too common paradox of my generation, a generation that seems to find its financial security while on the move. "What's saved this congregation and given us new blood is Dow," Reverend Westphal had told me, "but the families who've moved to Beaver in the past ten years are not necessarily going to be here a hundred years from now." A prominent newcomer, on the verge of being chosen the congregation's chairman, had been transferred to another job in another state. Yet where was the Kohn family going to be in a hundred years, or ten years, or even five? Where were the Ittners or the Endlines or the Gerstackers or the Voghtmans going to be? Who in Beaver anymore knew the future? A huge shift was occurring here, almost as if God were rearranging the landscape, and no one would be left standing in the same spot. "I felt like I'd woke up and found myself on Mars," Jean had said about her father's decision to give up the Ittner farm, and that, to greater and lesser degrees, pretty well summed up feelings inside the pioneer families who had moved—or been moved—off the land.

"Time for the big giveaway!" Chores done, my father clasped two of his grandchildren, Renee and John, and carried them to the Christmas tree. He randomly picked up two presents. "I'll read off the names, and you pass them out. See, you're Santa's helpers, and I'm Santa."

"You can't be Santa. You're not old enough!" Renee said. "You don't have white whiskers."

He liked that.

After the presents, and after a Christmas drink—one shot of whiskey—we assembled for photographs, first the full three-generational complement of twenty-three, then lesser subgroups, and finally the patriarch and matriarch by themselves. They sat side by side, without touching, on metal folding chairs. My mother folded her strong workaday hands in her lap. To the gold wedding band that was permanently on display, she had added today her special birthstone ring. She tugged her skirt over her knees for modesty. The skirt and matching jacket, black with a pattern of tiny red and white flowers, were out of style, but only by a few years. It was her outfit from the thirty-fifth anniversary celebration. Many times in those thirty-five years the most fashionable item in my mother's wardrobe, or my father's, was at least ten years old. He was now wearing a blue tie and a burgundy sweater, Christmas presents, about ten minutes old. He was sitting still, but not easy, his hands on his knees. The backs of his hands—burned by cinders in the smithy and almost crushed one time by a toppling anvil and abraded by tools, by willow brush and barbed wire, and by angry dogs—had a cross-stitch of white scars against the perpetual tan. The skin at the top of his palms was thick and shiny with callus, the thumbnails coarse as oyster shells, the fingers wide and unfrivolous and too large for his wedding ring, which my mother kept for him, hidden away.

During Fred Kohn's four and a half years in the Army, Clara Buchhage wrote him more than a thousand letters, almost one a day, and, from boot camp, from troop ships, from the front lines, he wrote back, although he could never hope to keep up. They had had to operate on the assumption that nothing was certain, and, in their letters at least, that uncertainty did not faze them. Their letters were mundane with the events of their days. Not love, not any feeling, was quite stated. Romance proceeded in rational imperturbable stages. After three years Clara was the only one of the girls Fred left behind who was still pressing her case, and he began to feel he knew her better than if he was courting her from the Kohn farm and seeing her every Sunday. He did not once bring up marriage, nonetheless. "Not in a rush to get anywhere," "happy with their lot in life," "good solid folks with their feet on the ground" are terms that friends used then as now for Fred and Clara. Finally, in August of 1945, he was discharged, and a few weeks later he proposed. And what of her answer? You can see plainly in photographs of Clara, her face wide open and festive, that life in Bay City had liberated her from the bleakness of the Buchhage farm, and, after all those years of honest wages from Dr. Kerr, she was financially independent. One could wonder, with her diligently saved nest egg, her wholesome prettiness, the rudeness of her departure from her girlhood in the country, why she did not marry in the city, why she waited around to become a farmer's wife. ("I didn't have a moment's hesitation. I didn't go out with other boys through that whole time. All I worried about is that he'd get home alive.") In October of 1945 Fred and Clara had announced their engagement.

Near the Christmas tree, the cameras of their six children snapped and snapped. Sitting, everything about our father was stiff, the stiffness you see in a horse before he bolts.

"Put your arm around Mother," Sandra suggested to him.

Ignoring her, or not hearing, he did not move. Sandra took his arm and moved it for him. Our cameras snapped. "I'm getting tired of sitting," he complained, and Sandra hushed him. He sat another minute, gripping the back of my mother's chair as if it was going to launch itself. "You through now? Okay? Good!" The cameras fell silent.

In my memory my father had kissed my mother perhaps once or twice in our presence. The only time that stood out, actually, was at their thirty-fifth anniversary. We had banged spoons and forks on the wooden foldup tables, as everybody does at German weddings, and refused to stop until they kissed. Now, with our cameras laid aside and our attention roaming, my father suddenly leaned over and kissed my mother, touching her with a special grace I had never seen. The effect on my mother was immediate; splashes of color sprang into her cheeks. Sandra whooped, "Hey, no fair. We want a picture of that." But Christmas Day had ended.

....

Ittner's elevator the next morning was a silver-gray city of tin silos and tin firebell cupolas and tin-roofed depots hung with snow. The scene faded into a gray sky. It was lost in front of me, an illusion of the early morning when sunlight comes through mist, off the tin, into mist again, and the effect is of a television set in which the picture tube is failing. To sketch it one would need gray pulpy paper with gray ink to outline it and silver paint for color. The illusion lasted into the parking lot.

"How's business?"

"You see any business?" Jean said stoutly, without any show of real worry. The middle of winter is typically a dead time of low prices at country elevators. Conventional wisdom says that whatever crops are not sold directly from the field in autumn should be held in storage until late winter, when prices should move up again. The morning line on navy beans, corn, wheat and soybeans, cued by computer from the Chicago Board of Trade, was posted in white numerals on a black adhesive board.

Jean was at the front counter in a building of wood and toasted brick that had a periphery of glassed-in offices squared around a bullpen of desks. The building was new. It was Oscar Ittner's declaration that his elevator would be here when these hard times were over. Having survived the worst—the Depression—the valley farmers could survive the 1980s.

Oscar was nowhere to be seen, though. "I'm letting Dad play hooky today," Jean said. He was resting up for a battery of special medical tests. "He's supposed to go to Lansing and get wired up like an astronaut." It was possible the doctors would recommend heart surgery.

Oscar Ittner was known in Beaver as a man of great decency and of shrewd diplomacy in his attention to the work of his neighbors, giving compliments where due and otherwise being silent. For several years he was the Beaver Township treasurer, and at Zion he had a record of service second to none. He was known as a crosser of T's and a dotter of I's. Many evenings a light stayed on at his desk, with him poised there, while his employees were already done with supper. Roy, who, just out of high school, worked two years at the elevator and almost made his career there, had found Oscar scrupulously fair and willing to work alongside his crew, round the clock during the peak weeks of harvest. A lot of stress is involved in being an ideal boss, a point his doctors were trying to impress on Oscar. ‘‘They want him to be less involved and take more vacations,’’ Jean said.

‘‘Tell him to come to Washington and see the sights,’’ I said. ‘‘I give a great twenty-five-cent tour.’’ I wasn’t speaking idly. Over the generations the Kohns and Ittners had been neighbors and friends. When Fred Kohn had five sons and Oscar Ittner had four daughters of comparable age there was speculation and wishful thinking, and, although no marriages came of it, some of us were smitten for a while. And no one would have been a finer father-in-law than Oscar. He had the bold entrepeneurial streak that ran in the Ittner family. His grandfather, J. P. Ittner, who immigrated to Beaver Township with Heinrich Kohn, was, at the onset of the 1900s, Beaver’s leading citizen. If the township had a capital then, it was Ittner’s Corners, and its main enterprise was Ittner’s sawmill. Along with Heinrich Kohn, J. P. Ittner was a founding father of Zion, and the congregation conferred on him the title of honorary elder. ‘‘The church minutes show his name occurs in all important projects,’’ the fiftieth anniversary committee reported, although by then he had lost almost everything in the steam-boiler accident that exploded his sawmill and in a stock swindle that wiped out much of the family fortune. But the Ittners had a farm to fall back on, one of the best-looking in the township, with sun-reflecting sheds and silos, a German barn, a white, shuttered frame house and a roll of fields that played a trick on their scale. They were in a wide, treeless rectangle—J.P. had saved no woodlot—and had the long-distance breadth of Iowa or Kansas. Oscar always spoke of his years on the Ittner farm as a lucky time. But Oscar also had a larger view of the world, and, even while farming, his style was Main Street, as if beneath his sweated work duds there was a starched shirt and knotted tie. It was Oscar’s businesslike appearance and attitude that called him to the attention of the elevator’s former owners in 1968. They were on a kind of talent hunt, scouring the area from Bay City to Midland for a suitable successor. (‘‘They drove around and looked at farms to see who would be a good manager.

They wanted someone local to take over, no outsiders.") Their reasoning was logically provincial: a buyer from outside the valley might use the elevator for a tax loss, letting it run down, or might cut off credit to farmers, or might have some other foreign approach, and thereby ruin the lifelong reputations of the owners. On Garfield Road, they saw Oscar's formidably ordered farm shining in the sun. They inquired around. Oscar had installed a grain dryer at his place, which saved him the drying fee when he sold his crops, and which, in addition, he rented to his neighbors. The elevator owners liked his eye for the thriving edge of the farm economy. ("Their offer came out of the clear blue. They presented very fair terms and said, 'It'll be good for you and good for the community.' ") Could Oscar keep his farm while managing an elevator? That was out of the question, he decided, and he made the choice he believed best for his family. With his choice, he rose in my estimation. Clearly he had inhaled a strong puff of Yankee ambition. When Jean came home from Germany, over her first Sunday dinner with Oscar, he began citing tons instead of bushels, and she knew he had moved on to bigger things.

Now, while Jean looked for an ITTNER BEAN & GRAIN calendar, a man opened the door and came up to me at the front counter. He was squinting and wincing. Outside, the mist had burned off, and it had turned into the kind of day when sun and road grit and little cannonball pieces of snow are swirled together and you are smart to wear sunglasses. He wore striped overalls, a cap with earflaps, orange canvas gloves. I knew the man from years ago. A World War II wound, of which he was profoundly proud, threw his walk off. He glanced at the price board. Stained teeth flashed into view as he spoke. "Might as well sell the rest of my soybeans. It's only going to go from bad to worse." Focusing on me, he repeated himself. "From bad to worse, wait and see. You can forget those rose-colored predictions the government's handing out. Who're they trying to kid?" The man made a face that was tangled and clownish. I smelled morning whiskey on his breath.

"Been rough, the last few years," I said, trying to be affable.

"The government is culling us out. That's their plan. Been their plan ever since the war." The man I remembered was a tall, athletic figure, someone who strode with long steps, kicking out his shrapnel-hit leg to give his stride more emphasis. Now he was a Wordsworthian figure, bent forward at the shoulders into the headwind of age, and he walked uncertainly. Leaning on the counter, he said, "Aren't you one of Fred's Kohn's boys? How's your dad making out? Heard he quit that deal he had with Don Rueger. Fixing to retire, is he?"

"No, he's got a new deal."

"Hard man to keep down! I don't know, though. This is no game for old guys. Your dad's five, six years older than me, and I'll tell you, I'm thinking of selling my place next year."

Jean returned with a calendar. "Here you are, compliments of the management." She beamed. "Tell Roy the guys want him back here if the nuclear plant lays him off. Oh, and say good-bye to Dale for me. His wife seems very nice." The last chance for a Kohn-Ittner wedding had been Dale and Jean's youngest sister, and it might have happened, except Dale left to see the world with the Air Force and never returned.

....

Preparing for Tri-City airport, Harvey zipped into a tote bag an orange round of cheese, a mild Cheddar, the cheese we grew up with. He had purchased it in Linwood, near the farm, at the Hitz cheese factory where, at one time, my father's milk cans were delivered in a daily truck run. At other times his milk cans went to Raber's cheese factory in Willard and to Swan's Dairy in Saginaw and to Kraft's cheese factory in Pinconning as his veal calves and butcher-ready hogs used to go to Peet Packing Company in Bay City (where it was hard to watch the knockout men with their sledge hammers and the skinners with their red knives, and impossible not to), and as his eggs used to go to the big IGA store in Kawkawlin. Almost nothing now was left of this infrastructure, the cheese factories and creameries and slaughterhouses and egg-and-vegetable markets closed or drastically reduced because of the new economics. With the last of the important closings, Kraft in 1978, the era in which town and country existed as equals in the valley, each dependent on the other, came to a close too, and a few days later my father sold his cows. Today, in valley stores, the eggs on sale might be from chicken factories in the Dakotas, the vegetables from truck farms in California or Texas, the pork chops from hog factories in Iowa, the beefsteaks from feedlots in Nebraska, and the cheese from factories in Wisconsin, which was the case with the Cheddar in Harvey's bag. No cheese came out of the vats at the Hitz cheese factory anymore either. They sat empty and unused. The Hitz family had given up on that part of the business and instead sold cheeses imported from elsewhere.

At the airport curb, saying good-bye to Harvey and Dale, my father gave them a stiff, uncomfortable handshake. "Don't try to do everything yourself," Dale said. "You've got a couple of sons who're still around to help out."

"Don't worry, I got the easiest job in the world. Nobody can tell me when to go to work or when to quit." He coughed wetly and cleared his throat. I thought he had finished. "It was okay to have you kids on the farm when you were little. You could do things right. Now I would have to teach you all over again." This was his way of saying it was okay for Harvey and Dale to be leaving after only a few days.

The Christmas reunion had succeeded on many counts. Dale had said to me, after Christmas dinner, "It feels like we're finally a family, sitting here and talking to each other." He and I had been born ten years apart, and, big brother and baby brother, we had not gotten to know each other until we were grownups. Stories exchanged with the holiday gifts helped narrow some of the gaps. The reunion had been filled with storytelling and long talks about sports and religion and politics, even one attempt over dessert to discuss the unknown future, that feared day when our father's heart might seize up. We eased into the discussion. It was uncomfortable and odd. The patriarch did not avail himself of the family grouped at the table to declare his own wishes. Did he want us to establish a long-term cropsharing plan with Gerry Radke and keep the farm in the Kohn name? What if we sold it? Or did he have hidden hopes that one of us would take his place? We did the talking, not talking to him so much as in his presence. He did not so much as lift his head, just sat and scooped ice cream into his mouth like a kid on his birthday. He seemed nonchalant, inhabiting that barricaded solitude of his. Finished with his ice cream, he looked down toward his idle spoon and pulled his mouth into a wryness that automatically meant the discussion was over. He held up a glass of water as if presenting a toast, and he became alert, interested. ''Tastes pretty good, doesn't it?'' he said, waiting for everyone to agree. He drained the glass. ''Yes, sir, purified water, just like they drink in the city.''

We did not return to our discussion. ''Never time enough for everything,'' Harvey said, getting on the plane. There had been gallivanting to see old friends, the side trip to buy cheese, more shopping for scarves and fur-lined gloves—''Either my blood's thinned out or the Michigan winter has gotten worse''—and hassling with the kids in the undiscipline of the road. On and on. Time had run out, although, of course, we could have made time. But we were satisfied. After daring to peek at the box of future deaths and hard choices, we had quickly shut it. Diana said later that she was struck by how the five sons, and Sandra as well, were unmistakably Kohns, eerie re-creations of our father, down to the tunneled-in eyes, the knuckly cow-milking hands, the strong silences, the closed-off feelings. And the false optimism. We had decided there was no urgency, because the day of ultimate decision was so far off. It was a kind of American rationalism, somehow pragmatic. We had a profound sense of relief, I think. Like other families, we were skittish about what might shoot out of that box. So we changed the subject to something with no psychic terror—to sports and religion and politics.

About sports, there was this: Sparky Anderson, the Detroit Tigers manager, was bragging that next year's team would be his best ever, a fair compliment from the man who had made his name with the Big Red Machine in Cincinnati. But the next season did hold immense prospects for the Tigers. It always did in December. Or almost always. I had grown up in the hated era of predestined New York Yankee championships, listening from my milk stools to Tiger games on WSGW. These days the Tigers had a real shot and did not usually slip from reasonable hope in the standings until late in the season. Loyalty to the Tigers had stayed with all of us. We were fans for life. All season long, at a certain point in the day, we searched out the Tiger box score in whichever newspaper we read, and whenever we talked on the phone we knew who was yesterday's hero and tomorrow's pitcher.

About religion, this: Roy predicted that Beaver Zion men would soon see the light of women's suffrage. A shortage of men willing to take on official congregational duties, he said, was about to undo tradition. Reverend Westphal, too, saw that Zion must move into the flow of liberation within Lutheranism. "In past generations, wives would say their piece to their husbands, and their husbands would bring it to the voters meetings," he had said. "Today's generation of women want to come and say it themselves"—Sandra among them. But someone who had spoken up emphatically for suffrage—my father—said to Roy, "He's got another think coming. They'll dig up men from the cemetery to be church officers before they'll let women in." I knew the congregation well enough to understand my father's pessimism. Far more of a mystery was his own view of Lutheran women, a 1980s view. He said it was based on the Bible. At a synodical convention once, arguing for the ordination of women, he had waxed theological: "The Bible was in German when I went to school, so maybe I read it wrong, but I seem to remember that God commanded for woman to be a helpmate, and what is a minister? Isn't a minister supposed to serve and nurture a congregation and be a helpmate? Maybe God meant for women to be ministers instead of men!" A radical, modern view—which, frankly, I attributed to Johanna.

And about politics, this: My father had voted for George Wallace, George McGovern, and Jimmy Carter, even though they were stuck with cultural separateness and were bound to lose. He liked them for being outsiders and for standing for regular Americans. He could scarcely believe that Carter, perceived as a farmer, had won. As for Ronald Reagan, my father had voted for him, but he was disillusioned about the trillion-dollar deficits, casued in great part, he thought, by Pentagon sloppiness and indulgence with a dollar, to which Dale took some exception. "Sure, there's waste," Dale said, "but the real cost is the red tape and paperwork, and Reagan's trying to fix that." Dale had tied his fortunes to the military, first with the Air Force, now as a computer technician for Singer Systems Inc., in which job he fine-tuned the simulation computers for the supersonic F-16 fighter. So much had changed since World War II, my father said. Back then the Army was a calling. "The guys I served with were what you'd call average joes, nobody bucking to be a general or get rich off the war. Oh, maybe a few were, but nowadays everyone is. These young guys sign up so they can learn computers or what not and hook on with the defense industry when they get out. I don't blame them, don't get me wrong. They're just going along with the way the Army advertises itself on TV." My father's argument with Reagan and the Pentagon and the whole of the federal government was that the current administration was widening the circle of people who depend on the constancy of government dollars. For that matter, it was also his argument with modern farming.

....

I have a photograph of four generations of Kohn women and girls. As far as I know, it is one of a kind. It was taken at the farm during Christmas of 1971. I was married to Kathy then. Liz, who was then "Elizabeth," was three years old. They are in the photograph, along with my mother and Sandra, Margo,

Carolyn—and Johanna. Only Liz and Sandra were born into their place; everyone else had married in. Yet in her long print dress, her heavy-mesh nylons, her old ladies' shoes, and with her enormous survivor's will, Johanna was more of a Kohn, I thought, than anyone. I had held onto the photograph even after awakening to the possibility of divorce when it was hard to look at the gay holiday smiles of Kathy and Liz. What rivets me every time I look at it is Johanna's smile, so absolutely wonderful: she was so undaunted by the years and yet so full of them, so in love with the familiarity and the easiness of everything even with her crippled hips, so delighted to sit on her cane-latticed highback, which made do as a sedan chair that we grabbed and lifted into position at the dining table. Her high, bright smile conveys to me that it was possible to be very happy on the farm.

This Christmas, before Johanna was taken to the hospital, I had hoped to update the photograph, Great-Grandma in with her new in-laws and her new great-granddaughters. My disappointment was great, in spite of the silver lining that her doctors saw. They thought her pneumonia may have saved her from the dangerous charge of a family reunion, after which there is always the chance of a letdown leading to disaster. A rather common denominator in the deaths of old people is that they come close upon holidays or birthdays or other special events. Certain old people, either very clever or very instinctive, try to protect themselves from such excitement; they will themselves sick beforehand and recover in short order afterward—so goes a theory anyway.

I knew in Johanna's case the theory was wrong. She had been looking forward to Christmas at the farm. In November, in her wheelchair at Colonial, she had said, "Next time I see you I'll be at home," speaking faintly but with an acuity that couldn't be missed.

I mention the theory only because a few days after Christmas, amazingly, Johanna was back at Colonial, little the worse for her stay in the hospital.

....

In sunlight sharp and white, my father and I took a pickup load of hundred-pound burlap-bagged navy beans to Ittner's. We lifted them by hand onto the truck bed and wedged them like pickles in a jar. At the elevator, the atmosphere was early morning: no one was in line ahead of us. One of the pit boys motioned us to stop. We were in a drive-through depot of rough lumber and open timbers. The pit boy, wearing a black snowmobile suit and a white nose-mouth mask that hung below his chin, dropped the tailgate and began to untie the bags. The truck was stopped a foot beyond iron grates that overlay a cavernous space bright with sheet metal. This was a pit. We emptied the bags through the grating, each a small landslide. The pit began to fill. The pit boy pulled his mask over his nose, expecting dust to come off the beans. None came. He reached out and grabbed a handful. A cull or two, a few splits—virtually everything else in his palm was oblong, white perfection. The pit boy, a teenager, whistled. "I've never seen beans come in so clean. What'd you do? Fan them yourself?" He was confused. It was his first winter at the elevator. He did not know about older farmers, who, pride on the line, an eye on top dollar, bring in crops already run through a fanning mill.

As many farmers do, my father sold part of his crop at harvest and stored the rest into the new year, hedging both ways against a swing in prices. When you keep beans in a granary, though, you must remove moisture or risk mold. The moisture is in green pods and unripe beans and in the stems and leaves of weeds. Don Rueger's deluxe combine sorts out this junk, and he feels he doesn't need additional fanning.

The oscillators and screens on my father's combine, an eight-foot Massey-Ferguson, were beat-up and sorted poorly, and he had to fan his beans. He would have done so regardless. He always had. The long history of his Clipper No. 1B fanning mill was expressible in trillions of hand-turned revolutions. The principle technique of a fanning mill is to shake and blow and move the ingredients across tiers of screens. My father was sophisticated with the old Clipper. He would dump a bushel basket of navy beans into the Clipper's hopper and, whirling the handle, would separate out four piles: stems and pods and dirt, for garden mulch; weed seeds and chaff, for the chickens; culls and splits, which, when he was dairying, he baked in a huge iron kettle and fed warm to the cows; and, at the bottom of the Clipper, in a red wooden box—the point of it all—the best of his crop.

"No kidding," the pit boy exclaimed again, watching the last of our load go into the pit, "these beans are beauts, really clean."

My father nodded, acknowledging the praise.

"C'mon, get going," the boy said to the conveyor belt at the bottom of the pit. The belt was rubber, and all along it, bolted on, face-up, were metal cups. "Damn belt's been sticking." The belt made a racket. It moved. The beans departed in the cups, which hold four pounds apiece, ascending into an internal railway, out of sight. The first stop was the upper bin, and from there the beans fell through a downspout into the big three-tier mill where a jigger blew out chaff and three electric eyes spotted moldy beans and ones of inferior size. From the mill, they went up a leg to the scale hopper, down a leg to the basement, back up a leg and through another downspout. In the legs were augers. Every load took the same route until the final downspouts, of which there were eighteen for the eighteen tanks. On a platform above the grating, the pit boy moved a sprocket wheel that had numbers printed in a circle, corresponding to the tanks. He stopped the wheel at number eleven, a navy-bean tank. In this season of no particular harvest, all crop possibilities had to be contended with, and the pit had to be empty of one crop before another crop could be unloaded.

"Here's your sample," the pit boy said. He handed my father

a coffee can filled with beans that had passed through the mill. My father probed the can with two fingers, rolling beans up from the bottom of the can. "Ought to pick five or six," he said.

The "pick" is decided by the percentage of fungal spots and mold and dirt specks, and also nicks, shriveling, swelling, anything that lowers quality. A pick of one is the top grade and forty is failure. The higher the number, the lower the farmer's take-home check. The elevator manager or a designated assistant arrives at the "pick" from a visual inspection, and it is as objective and as subjective as an editor's opinion. No matter how often you go through this, as farmer, as writer, it is seldom matter-of-fact. The workman in you is hung out. It is the moment when the labor is complete, the product has been submitted for appraisal and payment, and you take whatever judgment you get. This moment, as I've tried to explain to Diana, can be a flash of full anxiety—panic, if you will—when nothing at all stands between you and the disappointment you are forever imagining. It can be keener than any disappointment in school or in sports or with sex, more at the crux of individual survival than a winter storm or a bad heart. It is a venture of the inner self into the outside world, a time of truth, a naked exposure if not a baring of soul, and, when your work is given a top mark, a shine will extend through your next exile in the fields or at the typewriter.

My father carried the coffee can into the elevator headquarters. Jean was at the front counter. "Is Oscar here today?" my father asked.

"I'm here," Oscar said, joining us. He took the coffee can and spilled the beans into a tray.

During Oscar's first years at the elevator he found the "pick" an uncomfortably public and direct way of telling his friends what kind of job they were doing on their farms. Those days he had softhearted lapses. Now, with the market in its fourth year of decline, he had to be tougher.

This was a pretty easy call. "Only a few bad ones," he said. "It picks six."

"That's about what I thought," my father said.

"Whoops. Hold it. I forgot to test for moisture." Oscar returned the beans to the coffee can and inserted a specialized barometer that registers the water that the beans are retaining. He waited a few minutes and pulled it out. "Yeah, six. They felt dry. We had a load of soybeans the other day with snow blowed on. They tested thirty-five."

"Wind blow the tarps off?"

"No, he didn't think he needed tarps because he had them in a gravity-box in a barn, but the snow came through the walls. A thirty-five pick. I felt bad about that."

"I had to shovel snow out of the north end of my barn last week where it drifted in."

"There's something wrong somewhere with this weather. I can't remember a winter that was so fierce."

"Sputnik, that's what did it. All that junk up in space. The Lord gave man dominion over the earth, not the heavens."

"Maybe you're right."

Oscar rubbed circulation into his hands and began to fill out the "pick" slip. After fifteen years at the elevator, he had lost his outdoor face, the white shaded forehead and the sunburned, windburned ears and neck. His head now was a smooth, consistent, indoor pink-brown in every direction, hair gone almost entirely from it. He had a short, tipped knuckle of bone for a nose. His eyes were pale and blue, the clearest of windows.

"Jean will figure out your check," he said.

On the price board, navy beans had been at $16.22 a hundredweight when we entered. "You've been making money while you talk," Jean said, peering at the computer readout from Chicago. It was up to $16.27.

"Hurry up, before it goes back down." My father made a playful gesture.

The elevator had attracted other farmers dressed in denim, the rural costume that in the 1960s became the uniform of a universal priesthood, seen on *Esquire* covers, heard in California songs, and now was high fashion in New York. One of the farmers, a man from Zion, asked about Johanna. My father said her recovery was remarkable. "You know, she's the oldest member of the congregation," he said with pride. In the bullpen, Jean punched a calculator. "Want me to check the price one more time?" she asked.

"I'll take what I got," my father said. "Nobody loves a greedy man."

....

On New Year's Eve, after church services, we had a party at Sandra and Mike's with pizza and red pop and rum in Coke and Sandra's homemade chocolates in bonbon molds, which was about as much razzmatazz as we ever had. The Orange Bowl parade was on the television. The younger children tugged at each other and teased each other. "You'll never stay awake till midnight," Sandra said, trying to calm them down.

"Who wants them awake?" Mike said. "C'mon, leave them be. Come upstairs and get piggy with us old folks."

I had in my hand my sixth bonbon. Heat was pouring from a Shenandoah woodburner that Mike had installed on asbestos tile at the end of the living room. Family portraits were on the wall, although, for the time being, they were wrapped in Christmas paper and crisscrossed with ribbons, a fanciful idea Sandra had picked up from Roy's wife, Lorie. On a shelf there was a framed photograph of Sandra's softball team in parade dress, the blue-and-gold uniforms of Blackhurst Realty of Midland. Blackhurst had been the logical choice for a realty company now that Sandra and Mike were putting up their house for sale.

"Will Blackhurst give you a sign for the lawn?" my mother asked.

"Sure."

"That's good." My mother's eyes had a soft blue acceptance that said she would not spoil the party with any spoken regrets. As it was, Sandra and Mike's departure was no longer immediate. Sandra had decided not to take the transfer to Ohio. "Too much upheaval for the kids, changing schools in the middle of the year.

And Mike would have to stay behind at Ferris; we'd be apart for weeks or months,'' she explained. ''No, I'm better off to stick it out here. This way I'll be here for another summer, in case anything should happen to Da—well, just in case. Besides I'll get to play ball for Blackhurst one last season.'' If Consumers Power gave up on the nuclear plant, she would find piecemeal work to tide her family over.

''Good thing you're not going to Ohio,'' my father said. ''These Midland workers would sure miss you on payday.'' It was Sandra's job to see to the processing of approximately five thousand weekly paychecks. The nuclear plant had a concentration of workers second in Midland only to Dow Chemical.

''Can you imagine the mob scene if they laid off everyone at once?'' Mike said. ''It'll take weeks to register everybody at the unemployment office. Midland has three or four clerks, that's it. They'll be overwhelmed. Midland's never had to deal with anything like this. It's not like Bay City.''

''I heard that the state will set up portable offices and send in unemployment clerks from Flint or Lansing to handle it,'' Sandra said. She stirred herself to look at the woodbox. ''Who forgot to bring in wood?'' Mike and I headed for the door, and she called after us. ''Check on the kids. I don't want them messing around outdoors.''

A black Labrador was the only thing howling in the yard. He threw himself against his chain, leaping for attention. Mike raised his hand and pantomimed hitting him with a switch, and the dog laid back his ears and lowered himself to the snow. ''Not fit for man nor beast out here,'' Mike said. He unchained the dog and put him in the garage. ''Hate to make a habit of this, but . . . '' The snow broke like white glass under our feet. Above us, the North Star was in the northeast, nearly seven hundred light-years away, shining with gases that burned long before Heinrich was born. The North Star that shone when Heinrich crossed the Atlantic had burned at the time of the Crusades, and the North Star seen on earth when Columbus sailed had been lit up in the age of Charlemagne. Mike was a devotee of science fiction; in one of his stories there existed a telescope with the ultimate in optics, and through it you could see across the curve of time, spying deep into the past, into the light of Creation, discovering there whether it was God's work or a big bang or whether matter is still being created, constant with our evolution, as galaxies rush apart. This last idea is known as the steady-state theory of the universe and originated in the 1940s with three men: Thomas Gold, now a Cornell University professor of astronomy (who also has a theory that most oil and gas is not from prehistoric life fossilized in sediment but from meteor-like materials buried with the forming earth, that is, from inorganic minerals rather than plants and animals); by Sir Hermann Bondi, a Viennese student interned during World War II as a civilian prisoner of war in Quebec (where he met Gold, another Viennese internee) and who rose to knighthood and the captaincy of science at Britain's Ministry of Defense; and by Fred Hoyle, the science-fiction writer. In their theory, Creation knows no special time, neither beginning nor end.

From a lean-to behind the garage, Mike and I loaded our arms with wood that Mike had hauled out from an uncle's spacious woods, riding old logging trails in his pickup. Back indoors, Diana was on the floor, legs flared behind her, close to the woodburner. None of the adults were far away, although up close the woodburner flushed your skin to match Sandra's collection of rose-tinted Depression glass. The comfort zone was from five to eight feet away.

Throughout the Saginaw Valley, you saw how houses had been modified to reduce heating bills—mounds of dirt, sometimes to the roof, creating virtual caves; solar panels on the full face of southern walls; and, most commonly, stovepipes through a roof that meant a woodburner below. In 1973 the crisis over Arab oil had made energy self-sufficiency into a virtue. In the valley, the Optimists and the Jaycees—Sandra and Mike were conspicuous members—spread the conservation word, and, at Zion, Proverbs Chapter 26, Verse 20 was recalled: "Where no wood is, there the fire goeth out." Commercials appeared on television for cast-iron potbellied replicas of stoves that predated the picture tube. In 1972, when Johanna moved to the Colonial Rest Home, a single antique dealer had inquired about her old potbelly purely as an atavistic piece, but after 1973 there was considerable interest in it for its winter value (although, in truth, it was junk from extensive firings). Consumers Power had done its part for conservation, too, mailing out brochures that told of "energy robbers" inside your home and promoted the value of wall insulation. Now, in the Consumers Power management, there was a kind of frantic nostalgia for the preconservation years. The savings in winter fuels, valleywide, had been so tremendous and unforeseen that it accomplished more than Consumers Power bargained for. Homeowners no longer were a big constituency for the nuclear plant, which was another reason it probably would never go on line.

My father sounded a warning to Sandra. "I'd say you should get that "For Sale" sign out on the lawn as soon as you can—before there's five thousand houses on the market and you can't give yours away. The sooner the better, I'd say." That fatherism jarred, practical advice though it was.

....

✓

Two miles from the farm is the four-corner village that used to be Ittner's Corners, now called Willard. The story of Willard is an American story of sticking it out. In the 1920s and 1930s, Johanna's sister, Helen, and her husband, Emil Beiberich, ran a general store in Willard, with red pop and mosquito repellent and bologna pickled in a jar, a good living until the Depression, when they sold out and he lucked into a job at Dow Chemical. But their store was still there—Jack's Market—and, look behind the door, there was red pop in a cooler. Gottlieb Raber's cheese factory had been across the road. In Johanna's generation the cheese factory was a male preserve where girls, who came along with fathers or brothers on the morning delivery of milk cans from farm to factory, got to meet boys, somewhat older, often there by themselves, which defined them as men. Until I was ten I rode with Johann in his black Ford pickup on morning runs. Then the H. M. Schmidt Company of Saginaw, which had franchised Raber's factory, decided to close it for the sake of consolidation. But the cheese factory was turned into a heating and plumbing store. A service station with a mechanic on duty also was still in business, and so was the old German beer garden that had been renamed the Willard Hilton.

The Willard Hilton was a place of escapism, and after New Year's Day Diana and I ate supper there. The meals were a countryboard of quality. A *Midland Daily News* reporter, discovering the Hilton, had given it more publicity than it had ever known. "Dinner prices range from $5 to $9. Dress is, ah, well, casual—just don't tell anyone," he wrote. "As for the Hilton name? The story goes that when all the Dow executives began showing up in their big cars, the local people just decided the place needed a big name." It was a curious scene: chrome-heavy cars rutting on lawns without sidewalks, their owners dressed for a twenty-dollar meal, and, alongside, Beaver farmers jumping from their pickups to go in and hoist a few. What comment might have jumped from J. P. Ittner? Perhaps only the amused surprise that natives have for the exclamations of tourists.

Diana and I sat at a corner table and gave our order to a waitress. Two men were tending bar, Oscar Ittner's brother, Art, trying to keep busy now that his son had taken over his farm, and a cousin of mine, John Kohn, the Hilton manager, so young I had to introduce myself. Our meals were served, but I poked at my pork chops. "Okay," Diana said, "what's the matter? You've been like this all day."

"I guess it hit me that we're about to go back to Washington, and what have I accomplished here? Not much. The farm is still teetering on a precipice."

She frowned. "That's a little melodramatic. This coming year is all taken care of, even if it wasn't your doing. Or is that it? You didn't get your big chance to come to the rescue?"

"That's not fair," I said defensively. I fell quiet, brooding.

Diana offered me a bite of her steak, and I accepted it. "I'm sorry," she said. "I know you came here excited about your ideas for the farm. Do you feel I stopped you from discussing them?"

"No. They were out of the question. Government subsidies, a government grant—I don't know what I was thinking."

"I wondered about that myself. Your dad would sell the farm before he went on 'welfare.' "

"When I'm in Washington, I guess, my perspective is so different. I forget what he's like, what Beaver is like." I brooded some more. "I don't know. When I'm here everything is more personal. I start feeling that something should be done, and what am I doing? Not a thing. Not a damn thing!"

"Am I to read between the lines that you're beginning to feel you made a major mistake twenty years ago?" Diana squeezed my hand. "Is that it? Please say so if it is. Don't hold back on my account. Do you want to move back here?"

"It's more complicated than that," I said, wavering. I felt tongue-tied, taken aback by her point-blank question. It released feelings and sensations in me that until a month ago were unfamiliar and were still difficult to pinpoint. Professor Salamon, the Illinois anthropologist, had written that Germanic sons today "must trade off some ideal goals against the demands of the market and of family members (i.e. fathers)." But that only explained the obvious fix that the heirs of many farmers are in: Stay on the farm and go broke, or leave and feel bad. It did not explain me. I had left without guilt. But what, then, was this longing in me? And how could I make Diana understand in a few hackneyed sentences what I did not understand myself?

I was good at leaving things, and the proof was in a wife left behind and a faith left behind and friends and cities and jobs—and especially a daughter, who, after the divorce, when Liz was first enchanted with horses, had conceived of a dream summer, father and daughter at the farm, she with a horse pastured by the woods, I writing half a day and teaching her to ride the other half. I had more or less promised her such a summer, and once in a while she wistfully recalled it and her phantom horse, even after she had long since realized her father was too wound up in himself. Where in his schedule of travels and dinner parties and political events would he find a spare summer? Every trip I made to Michigan seemed to turn into yet another lost chance to make up to Liz for the months and years of separation. Even during this long Christmas holiday, she and I had managed very little time together. The dry realism was that I had left her behind, like everything else.

And nothing was farther behind me than the farm. I had fought it and left it. I was not a farmer. That was the driest of realisms, the plainest of facts. There was no riddle about it.

"I'm ready to go back to Washington," I said.

"When?"

"Tomorrow."

....

Dawn looked like a white bowl. "Going to snow again," my father said.

"We'll make it," I said. At eight-thirty Diana and I drove off. Our car was packed with Christmas gifts and jars of my mother's sauerkraut and dill pickles. South on Carter Road, my father followed our track in his pickup. At Midland Road he turned for the Colonial Rest Home. Johanna's brother, Henry Walter, a retired Bay City barber, donated haircuts to the Colonial residents on two or three mornings a month, and afterward he usually drove out to the farm for a couple of dozen eggs. Today my father was saving Uncle Henry the second half of his trip. Diana asked if I wanted to stop at Colonial for a quick hello and good-bye, but, with the whitening sky and the drive ahead, I said no.

FIVE

Dale is in Massachusetts doing computer repairs for the Air Force, spending our hardearned tax money, $3,000 a month plus salary. But, the way I hear it, The Air Force is happy to pay him. Without Dale, the Air Force would fall apart. Same with all you guys. Without Roy, Ittner's would fall apart. Without Harvey, Miller's would fall apart. Without Ronald, Helfrecht's would fall apart. Without you, Rolling Stone *would fall apart. Now I know why I didn't get a farmer out of any of my five boys.*

—from a letter, October 1977

It is a peculiarity of modern sons that so many of us want to outdo or undo the work of our fathers.

In 1962, my freshman year at Bay City Central, a report from the Committee for Economic Development was circulated in Middle America's school systems. Titled "An Adaptive Program for Agriculture," it called on educators to persuade farm children that a better life awaited them in cities. The authors of the report, highly regarded economists and sociologists, had an unconcealed agenda. In view of the Green Revolution then taking hold in the Farm Belt, they had concluded that the time was right for mass efficiency to prevail over sentiment. "Large numbers" of farm children should be "removed from agriculture before they are committed to it as a career." One million businesslike farmers could produce more than the four million who were still trying then to hang onto their land. The one million, by acquiring the holdings of the other three million, could begin to operate on a corporate level. The report was sponsored by Ford Motor Company, H. J. Heinz Company, American Telegraph and Telephone, IBM, General Motors, Standard Oil of New Jersey, the American Can Company and several other corporations, certain of which had a vested interest in corporate farming—Ford sold tractors, Heinz depended on the economy of scale in tomato plantations—and which, collectively, in the general course of doing business, found it more comfortable to deal on a corporation-to-corporation, executive-to-executive basis, and were interested in having agriculture come into the second half of the twentieth century. At the same time, three million farmers and their children could be liberated. James Bostic, a U.S. deputy secretary of agriculture in that period, said in a much quoted speech, "Just think what it means to a society when more than ninety-five percent of its people are freed from the drudgery of growing food." The report's basic premise was widely accepted. The children of the Farm Belt, whose numbers were reduced in the first half of the century by the Depression, by world wars and by the rigors of work, now came under pressure to choose an education that would get them off the land. "You test high. You have an aptitude for half a dozen professions," my counselor at Bay City Central told me. But I did not feel this as pressure; it was flattery. He was telling me what I wanted to hear.

An old farm, in my generation, was quite clearly no place for anyone who had anything on the ball. It was a place of the immigrant, an opportunity of nineteenth-century America. Heinrich had stood at a flash point of history in the Saginaw Valley—the opening up of the forests, when Germanic imagination ran from tree to tree, sawing them down to let the sun in. Johann had expanded the farm's open space and had advanced its production, and by dint of conservation and labor had beat back the Depression. In photographs you can see the pioneer in Heinrich and Johann's faces. They are remembered for aspects of self-sufficiency and grit. In their time, black bears were in nearby woods; the unpeeled logs of a lean-to had to be chinked with mud; their wives churned butter; Chippewas and Sauks lived in the community and walked in the woods along trails of curved branches that had been bent in trees when the trees were saplings. By my time, the bears and the log structures and the Indian trails were gone, and, for me, the romance. Farms are not for the romantic, of course, but rather for the pioneer, in the full modern sense of the word. The pioneering thing for my father's generation of farmers was to acquire more land and more technology so their sons might inherit a potential corporation. In May of 1967, a Bay City *Times* photographer snapped a picture of Ronald, unsmiling, clear-eyed, at the annual FFA banquet. He has in his hands his "outstanding senior" award; his hair is flipped up, and he is wearing a blue jacket, corded, his name scripted in gold thread, the FFA emblem. The award he lost track of, but the jacket he kept in a basement closet. He had brought it out when I was last at his house. "I sold *Farm Journal* subscriptions to buy it. Only wore it once or twice. I don't think it's ever been washed." He inspected it: "Nope, never has." I asked him why after four years in the FFA he had changed his mind about being a farmer. My question seemed to stir him and then dishearten him. "What was I going to do? I didn't have the bucks to buy my own place, and, after the deal for the Pashak place fell through, that was it," he said. The Pashak farm, eighty acres, adjoined the Kohn farm on the south. Albert and Viola Pashak, childless, afflicted by gout, hoping to sell and find an apartment in Bay City, had offered their farm to my father, but there was an impasse over the barn, and then there was the matter of the house, white with pink trim, smilingly out of place in among the Germanic plainness, Viola's idea, taken from a magazine. My father did not want it to appear he was pushing the Pashaks out of their house into the exile of Bay City—no matter that they were more than ready to leave. ("That old house, they couldn't wait to get rid of it," my mother said.) So the negotiating rights to the Pashak farm passed to another neighbor, Vern Chapman, who moved quickly. At Christmas, home from college, I saw the Pashak house, a knocked-down pile of boards and rafters, showing pink here and there. "What do I need another eighty acres for?" my father had said combatively. "I've got enough at home to worry about."

Later he had other opportunities to increase his acreage, and he had another son, Roy, who excelled in the FFA—a flaking plaque nailed to the toolshed wall reminded us of Roy's FFA status long after it was past—but the story repeated itself.

On meeting my father, friends of mine have wondered why he confined himself to one small, unimportant piece of land—exactly the question that ran through my mind when I was young. He had the brains to have reached the pages of *Successful Farming*. He was knowledgeable about federal trade policies, foreign markets, the tax code, in short, all the homework of the corporate farmer. Arithmetic for him was a snap; at Ittner's, he could figure out the amount of his take-home check ahead of the calculator. And his memory always stopped me short. At Christmas, happening across a group photograph of the forty-six boys and girls who attended Zion's grade school in 1924, he was able not only to recognize and recall everyone by sight but also tell me what they had done with their lives, where they had lived, which church they attended. He was dismayed that he could not quite remember all their spouses, but later he chalked that up to missing out on several weddings during the war. In the school photograph my father is a second-grader in long pants. He is big for his age, with a shock of hair. After the eighth grade he quit school to work full time on the farm, and for the longest time I thought his dropping out expressed a lack of ambition or a lack of talent. Either way, he was content with the status quo and therefore content to lose out. I did not understand Isaiah's warning, "Woe to those who add house to house and join field to field until everywhere belongs to them and they are the sole inhabitants of the land." I did not care that the world was rank with opportunism. I saw only the opportunity. The difference between my father's farm and the rest of the world was like a threat to me. I did not want to be trapped in Beaver. At Bay City Central, I caught the expansive, big-city enthusiasm. It happened that my eighth-grade class at Zion, fourteen strong, was the first to attend Central High, the school of moneyed Bay Cityans. Previously the students from farming districts had been tracked to another high school with an emphasis on vocational studies. In 1961, because of consolidation of school districts, the school buses took us to Central High from twenty and thirty miles away. We were not welcomed. A decline in academic standards was predicted. On my first day at Central High, preppy glamorous upperclassmen asked me where I was from, and I admitted with a touch of sheepishness, "Beaver." I was marked as a farm kid. The marking was done with lipstick. I scrubbed my face in a lavatory, and as soon as I walked out I was smeared again. The shirt my mother had ironed the evening before was a mess. This went on for days. We were taunted. Finally there was a climactic evening in the Mister Hot Dog parking lot on Euclid Avenue, in which we were represented best by Arnold Sanchez, the son of settling-in migrant farm workers (and later a Golden Gloves champion, killed in Vietnam) who, boogying forward, invited all comers to share the action, which to their chagrin they could not handle. Acceptance in other ways was eventually achieved, on the sports fields, in classrooms. Elmer Engel, the football coach, grasped what the coming of stout, hardened farm boys could mean, and in my sophomore year we won the Class A state football championship. In my senior year, at baccalaureate services, I gave a speech and said "Beaver" to applause; I was a one-two academic punch with Janet Mieske, also from Zion's little parochial school, the two of us graduating first and second in a class of nine hundred. I was awarded three scholarships to the University of Michigan, and ahead of me lay any number of choices and adventures. It had been an incentive that the city kids thought I was from a hick place. Yet—and this, I know, was at the heart of my departure from the farm—somewhere along the way I had bought their line.

But then my father had bought the line about me. He would tell a story, a sharp-edged family fable handed down by Heinrich, who adapted it from the New Testament. As the story goes, Heinrich went to Bay City one day and that night asked his young sons how they had used their time. Charley had milked the cows, Herman had fed the pigs, Johann had hoed in the fields, and to each of them Heinrich gave his blessing, "Well done, you will grow up to be a good farmer." But Henry had spent the day reading a book, and to him his father said sadly, "I guess you will end up going to school." As I grew up, I thought the story was a latent commentary on me.

....

After the Christmas reunion, back in Washington, Diana and I picked up life as we knew it. Within two days of unpacking our bags, they were packed again. Diana took a plane to San Francisco to work on a photojournalistic book about nuclear fission, and I went to New York for a series of television projects. We were apart most of a month. On the evening that Diana flew in from San Francisco, as she undressed, she said quietly, "I think I'm pregnant."

I was in bed. "You can't be serious," I said. We had never gone through the multitude of reproductive tests, but we assumed, after ten years of marriage, that we were not meant to have children.

"We'll know in the morning," she said. "I bought one of those do-it-yourself rabbit tests."

I opened the quilt and pulled it around us both. "Would you be a farmer's wife if I asked you?"

"I'm your wife."

"I need to do something about the farm," I said. "I'm the oldest son. I'm not supposed to let the neighbor's kid do it."

In the morning Diana's drugstore kit tested positive. I was in the shower. Diana, her voice catching, announced the news at the bathroom door. I grabbed the shower curtain, and, as if this somehow put me in direct touch with her, she moved to me in her nightgown. "This is going to be the best year we've ever had," I said. Diana braced herself on my shoulders, her eyes shining, and put an end to words.

The next two weeks we spent in New York, working and seeing friends. We rode the train back to Washington on a shabby, late-winter night and got to bed well past midnight. At 6 A.M., on a Wednesday, my father, who had phoned our empty house several times the previous day, made it through to ask me how soon we could be in Michigan.

"We can get on the road right away," I said.

Diana woke up, extending her arms from a nightmare.

"Shush," I said. "Everything's all right. Go back to sleep." While she did, sunk in forgetfulness, the quilt over her head, I began to pack the car. It was possible to extrapolate spring from the dawn. The eastern horizon, so long shrouded and dull, was lit up like an object freshly wiped. I woke Diana a second time and told her my grandmother was dead.

....

Johanna had died two months after her Christmas illness and recovery. The recovery had only been wishful; this was obvious to her nurse attendants at the Colonial Rest Home. Yet she was serene about it. Her room became her favorite place. The shade over the window was often pulled down, and the light in her room was a mixed yellow-white, alternately winter and summer. True summer light, searing light, once had been everything to her. At dawn, at its first seep, she was always awake, and within minutes always at work. On faint, grainy mornings she was sluggish. I don't think she was a Seasonal Affected Disorder victim, those people in whom a hormone runs amok in winter, throwing them into a twilight zone, but she did seem to have a hormone that energized her as days grew longer. One balmy April she planted such a large, repetitive garden that she had to seek out friends two townships away to get rid of the surplus. At the rest home, of course, her activities no longer were tethered to the sun.

On the Sunday afternoon of March 4, after the nondenominational service at Colonial, Johanna said to my father, ''I'm fine, Freddie. You take care. Wrap up good.'' On Tuesday morning, when he visited, her blood pressure and pulse rate were below normal, but there was something quite deceptive in her healthy color, and no one considered her an emergency case. She was napping. She stirred in her sleep but did not wake.

My father left to drive my mother to Zion; the World Relief had extended the quilting season. About one o'clock, while he was alone at the farm, a nurse from Colonial called. ''I'm sorry, Mr. Kohn. Your mother has passed away. She never woke up from her nap.''

I would have expected my father to kill time, to walk it off, to feed the chickens, and to wait until after the quilting bee to tell my mother. Instead he drove immediately back to the church. (''He was low, real low. I can't remember ever seeing him like that. He said they'd told him she went peacefully.'') That was the news Diana and I returned to. Arrangements for the funeral meal, the floral displays, the memorial fund, a gray steel casket with silver lining—''Everything's taken care of,'' my mother said firmly.

Inside a viewing room at the Cunningham-Taylor Funeral Home in Auburn, at the foot of the casket, in among wreaths on stilts, my father and his only brother, Martin, had seized floor space. They stood, dark-suited, facing a special traffic. Rough-handed men in clean clothes and women with winter-red faces and hushed voices presented themselves. They said very little. The style of German Lutheran grieving is set rituals, set language that leaves much of the grieving to the imagination. My father and Uncle Martin said, ''Thank you for coming.'' For a while my mother and Martin's wife, Erna, also stood in the greeting line, but, tiring, they sat against the wall on folding chairs with everyone else: the grandchildren; the great-grandchildren; Johanna's two sisters, Helen and Hulda, tired after plane rides from, respectively, Kansas and Nebraska; two brothers, Harold, up from Florida, and Henry, the retired barber from Bay City; nieces and nephews and younger cousins and neighbors present and past, although not many of those who had been drawn off to a wayfaring life in the cities. There was one young couple, who had never met Johanna but who came in place of a bedridden aunt as a final thank-you for help rendered by Johanna during the Depression. Almost all the inheritors were here in their inherited German Lutheran style, rooted in Teutonic tribal worship, incorporated by Martin Luther into a catechism and by German composers into down-tempo hymns, imposed in Michigan by Heinrich Kohn and J.P. Ittner and other church patriarchs: layer upon layer of history in with the day-old flowers. The low hum of music floated through the intercoms, above the quiet, shortened adult chatter. An undertaker at the door directed the old formality. It was as if we were in another world. But here the ritual was not borrowed; here it belonged; here the death of the very old still had meaning; the European Middle Ages still lived; here there was high reverence for the caste of family which once held things together against bad weather and sickness and prodigal sons.

Don Rueger approached the casket. He might have felt himself examined. The breakup of the Kohn-Rueger partnership was common knowledge. Don put out his hand. "Good to see you, Fred."

"Glad you could come."

"Hard to believe she's gone. She made it through all those other winters." Don paused at the head of the casket.

"This one was harder than most."

"Sure was."

Don's father, Harold, handed my father a memorial card with a donation for Zion. The similarity of the Rueger father and son—broad, nut-brown faces, high white foreheads, hearty voices, a look of bigness composed of many expansive muscles—stood out in the room. Don's mother, Thelma, was with them, and when the Rueger family was seated my mother made it a point to move and sit next to Thelma. That they had gotten along so well over the years was largely attributable to a sense that as farm wives their lot was thrown together and to plain talk between them. ''It's no good for anybody to be mad at anybody else,'' my mother had said, and Thelma agreed.

....

I awoke the next morning in darkness that had the quality of coming light. On the walls and ceiling and on the old, heavy furniture the morning began to spread. Small patterned flowers showed up like raindrops on the wallpaper. For all its fatigue, this bedroom, where once slept a young Johann and Johanna, then a young Fred and Clara, and now, on visits, Diana and I, gave off a spring freshness.

Last night I opened the window a crack for a breeze, but this morning the breeze was a hooting trickster March wind. Downstairs my mother had started a fire in the Kalamazoo. ''A cold front's coming down from Canada. It'll be winter by this afternoon,'' my father said. His voice was tight. He had the *Midland Daily News* open to the forecast. He turned the pages of the newspaper. ''Here—here's the obituary.'' He read it slowly. He had not slept well again. ''They did a good job. Spelled everyone's name right.'' A fierce, brittle look straightened his features, but when the brittleness left, his face sagged and became so wandering that I could see how someday he would look like the oldest pictures of Johann. He folded the newspaper, centering the obituary in a rough square, and, leaving it on his chair, went outside to the chickens.

My mother, stepping onto a stool, reached high into her cupboards for her best china. It had a pattern of wildflowers, and it had been a wedding gift. For the first several years of her marriage the china sat unused. The kitchen was Johanna's then, full of her things. I remember Johanna standing at the sink. There was room for only one person. Her back was turned, her hands active, in control. My mother had watched her, and I had watched my mother. The scene was dreamlike—the hot, confining room, a kind of domain, two women in whose features I could trace my own, one so sure of herself that I never doubted her right to be there, the other outwardly patient and complaisant but whose gaze revealed how much she felt an intruder. Over the years, slowly, my mother redid the kitchen and took it over, and at some point the roles had reversed.

But the old washboard sink was still in its same corner behind the door with the terry-cloth towel on a roller bar. The hot and cold faucets had been installed backward and had never been made right. From them uttered whiny, groaning ghosts. The sink was from a time that should have been gone and was like a denial of my mother still. Above the sink was a cabinet that held my father's medicine bottles and held, too, a physical, ulterior sense of foreboding: the bottles told the present and foretold the future. Everything in this corner of the kitchen felt haunted.

....

The pallbearers, the six of us, took our positions on the church steps. A peculiar stealth had overcome us. Johanna's heavy-looking gray casket was at the first landing on a long stainless-steel wagon, and we stood with expressionless faces, like soldiers about to be decorated. Then we lifted. The casket felt light as a woodbox. We stamped up the stairs, through Zion's tiled lobby, through the tall doors, down the aisle, past the smooth maple pews, to a central location near the baptismal font. The casket was opened. How waxy her face looked, worked over by the mortician's cosmetics; how fragile the neck, above the blue spring dress; how frail and quiet her folded hands that once had a loose, far-flung quality. We sat in the front pew, in a bath of light through stained glass. The organ stopped, and Reverend Westphal began to quote the Bilbe. "The Lord giveth, and the Lord taketh away." A warm retrograde feeling crept through me. Today I belonged in the congregation.

On the day Johanna Walter married Johann Kohn, in this church, she put on his finger his wedding ring, a thin piece of gold-plated brass mail-ordered from the Arbuckle Brothers Co., a coffee company. He had redeemed one hundred coffee coupons for the ring; nine cents covered postage and handling. Johann wore the ring on his wedding day and not again until he was buried. He kept it in a box lest it be damaged. Johanna and Johann were in their seventies when I began to see them, not as Grandma and Grandpa, but as a married couple. They had been moved out of the farmhouse to live by themselves on the back forty. I thought they would resent the move, and perhaps they did, but they made their new house—three rooms and a basement—into a honeymoon cottage, and in the period after Johann's first heart attack they were suddenly like young lovers, so bountifully attentive to each other. I learned I should knock rather than barge in the way I was used to. Nothing very shocking would be going on, maybe Johanna holding a mirror and guiding Johann's hand with sweet clumsiness so he could shave in his easy chair. But it was their place, and it was private, changed from the free-for-all of the farmhouse.

They planted fruit trees and Concord grape vines and berry patches and an overflowing vegetable garden. One day in the garden, after his death, Johanna said to me, "Sometimes I go out along the crick and back to the woods and I see Pa waving at me. When I get up close, he's just one of the trees. It's my eyes playing tricks."

I was embarrassed for her state of mind. "I think that sort of thing happens to a lot of people after someone dies," I said.

She refused my cover story. She said, "Even before he died, I saw him like that in the woods. I thought I saw him, I mean. He wasn't old anymore. He was a young whippersnapper, come to take me out." She looked to me perfectly calm, perfectly sane. Years later, reading about Zen teachings, I decided she had achieved a level of transcendence and was seeing with her "third eye." She had transcended old age, I thought. She had a girl's view again—which, on second thought, sounded like an old lady's mind wandering in time. It was Diana who helped me arrive at an intuitive explanation that satisfied me: "She was in love, and I think she was seeing what people in love see."

From descriptions I've heard of the young, dark-haired Johanna, she projected the kind of appeal that is usually associated, in the mythology of farm girls, with blondes. One of her suitors was Johann's younger brother, Henry, with a face for an oval picture frame, who involved her in a merry gang that went on hayrides and sleighrides. But Henry was restless, an ambitious and discontented young man for all his good-timing, and, one generation after Heinrich settled in Beaver, Henry became the first Kohn to leave. He went off to Bay City. He became a line supervisor at the World Star Knitting Mills. Evenings he trained for higher management at a business school. Except for Christmas, he kept a cool distance from what he might have called the common life of Beaver. It was an accepted fact he would make something of himself at World Star, built on the ashes of a sawmill on the Saginaw River east of the Third Street Bridge. Along the river, the twentieth century was underway. Newspaperman Leslie Arndt recorded the new sound: "The river was quiet. Lumber operations had all but ceased." Water Street was a dead end of broken-windowed saloons. A few years before it had been "Hell's Alley" where, the WPA history of Michigan says, "pleasure-bent" lumberjacks were entertained with whiskey, women, dancing and fighting. Several saloons had basement tunnels that led to the river—catacombs—"from the trapdoors of which dead brawlers and victims of skulduggery were sped into watery graves." Saloonkeepers trapped rats and let them loose in pits suspended below the main floor. A fast terrier, dropped into the pit, could snap the backs of five or six rats a minute. "A once-famous Water Street character sometimes took the terrier's place in the pit and was accordingly nicknamed 'Paddy the Dog.' " Bay City was of men and for men, who outnumbered women ten to one in 1900. Lumberjacks worked and played with violent exertion, and, since that was neither desirable nor easily convertible to other lines of business, they were the sort of men who, sometimes maimed, sometimes alcoholic, ended up on doorstoops and had no future whatsoever in Bay City after 1900. The new century belonged to new arrivals: farm boys from the outlying townships—Henry Kohn from Beaver—and farm girls. Henry's crew on the sewing machines at World Star had girls he knew from Beaver. Johanna's sister Hulda was on the crew. Upstairs at World Star was a roller arena, a safe, reputable hangout for girls; my mother hung out there a generation later. Whereas a farm boy was prepped and primed to follow in his father's steps, a farm girl was free—and often pushed—to make her own way, and from Beaver many teenage girls left for Bay City seeking either a nest-egg dowry or a husband right off. Bay City debutantes were known to say that a farm girl was an escape mechanism by which Bay City bachelors could remain boys a while longer. The marriage quotient in urban-rural romances was not, to say the least, one hundred percent. Hulda had to move on to Detroit before she met and married Claude Skinner, an agent for the Frankenmuth Mutual Auto Insurance Company. Johanna's other sister, Helen, worked as a housekeeper in Bay City and found Emil Beiberich, a traveler from Ohio, twice widowed, who opened the general store in Willard, selling to farmers a panoply of American goods. Martha Kohn, who was Henry and Johann's sister, married Carl Arnold in Bay City. He established a dairy, buying and selling fresh milk and processing butter and ice cream. But Martha's sister, Hannah, returned to the Kohn farm a spinster, and did not marry until later in life.

Johanna also worked in Bay City among Henry and his new friends, and she could have married any number of city boys, so it was said. But she came back to Johann. In my pew, I tried to imagine the moment in her life, the deciding instance, that turned her away from Henry and the city and committed her to Johann and the farm. Was it when Johann bounced onto her porch with the news that he had enough savings to quit her grandfather's sawmill and to begin to buy his father's land? Was it when he planted across the front of the Kohn house the maple and ash saplings that he knew would not be shade trees until he was a much older man? What was there about Johann that drew her back to Beaver? His work habits and steadiness? His faith in the future, or, put another way, his faith that his father's past could be his future? His rootedness? His undeterred approach to all he did? His undeterred courtship of her? ("He'd do his evening chores and then come in his buggy to see me in town, twenty miles away, and he'd have to be back in time for morning chores. 'Course he was young and didn't need to sleep.") Or was there something in herself, some unarticulated, half-mysterious homesickness for the farming life? One time I thought she was right on the lip of explaining her decision. We were in her garden alongside the little house on the back forty. Her hair was brushed smoothly under a gray pin-striped cap that matched her coveralls. On her thin wrist was a bracelet that had a gold locket with a picture of Johann. One minute her manner would be ladylike, reposed, almost delicate, the next she would be wielding a hoe with undisguised vigor and frank enjoyment. We were on the slant of a rise, an exposed hillside, from the top of which we could see Zion's church and school, and below us the green luster of the farm, scenery that she contemplated a few minutes. There is a beauty that is for me the beauty of life, and it flashed from her eyes, from platinum-pewter orbs and torn-looking gray flotsam caught in her irises. "I'd sneak out to play sometimes when I was in school and I'd come here," she said. "I remember being right here, right near where we're standing, and I'd think 'Someday this will be his' "—she meant Johann, who'd dropped out in the third grade and who may have waved back from the fields while she played hooky on the hill—"and someday it'll be mine." There was a half-beat delay, and she corrected herself. "No, I couldn't have known, could I?"

Now, Reverend Westphal leaned into the light of the pulpit and said loudly, startling me, "What can we learn from Grandma Kohn? Well, one thing is how to anchor ourselves. She wasn't like some of us modern folks, perpetually dissatisfied, worrying about what we don't have instead of making do with what we do have. She had what counts, her home and her church. Those were the centers of her life." He turned up his volume. "That's what we can learn from her." This was spoken like the line before "Amen," but he was merely underlining his point. He went on, "The last few years, when it was my day to visit her at the rest home, I'd say to myself, 'Why go? What can I do for her?' But I'd go, because the Lord expected me to, and there she would be, always smiling, always generous with her concern, always happy, and I'd realize the Lord was working in mysterious ways again. I couldn't do much to make her life better, it's true, but every time I went she made me feel better." He was at the finish. "That's the lesson we would do well to learn from her. Amen."

In the pew opposite us, my father and mother stood, and Uncle Martin and Aunt Erna. While aspects of Johanna could be found in her two sons, and in her grandchildren, she had been a character all her own, and proportionately more vivid. Diana, when she first met her in the rest home, thought the marvel of Johanna was that her strong, vibrant smile had survived intact despite the hardships of her life—the death of a son within a week of his birth, an Auburn bank failure that cost her and Johann part of their savings, and hundreds of ups and downs. Someone as enduringly happy as Johanna, Diana said, must have been born that way. I had thought so, too, but this morning before the funeral Johanna's sister Helen, reminiscing, had said that although Johanna was a happy-go-lucky girl, she did not become truly and noticeably grounded in happiness until she was a farm wife. ("On her wedding day, she said, 'I feel like a queen on a throne,'and from then on I don't think there was a day when she didn't feel like that.") Her happiness had come with her choice, as I understood it, and, until she made that choice, she had struggled against temptations from without and against elements within. ("Oh, she hated chores and wanted to run off, like we all did. And she enjoyed the boys and pretty things, dressing nice, smelling nice, going out on the town.") The true marvel of her, it seemed, was that she had chosen the farm after experiencing and liking the city and had not afterward expressed a single regret. Rather than grinding her down, the farm lifted her with little touches of fun, touches of hope, and its everlasting satisfaction. Every morning, up with the sun, her life was primary and constant, rising above all the varying, stressful events of the day. Her energy had been larger than life, memorable even to those who met her once or twice, and it faded only after she broke her hip the second time. "The Lord must want me to take it easy," she had said, and had smiled, turning her positive face toward the world. "At my age, I'm ready." One could draw a parallel to her at other ages, at any age.

We bowed our heads to pray.

....

Outside, I looked in vain for the bluebirds that sometimes return to Michigan in March and subsist on bugs stirring in air pockets beneath the snow. We pushed the casket into the hearse, and it pulled away, throwing gravel onto the snow line. A grave was ready in the frozen cemetery, dug by Roger Landowsky's backhoe, but the prayers at the gravesite were canceled. The wind was screaming with frightful, terrible power. "No use anybody else getting sick," my father said.

For eight years, across the road at Zion Lutheran elementary school, I had tried to learn the contradictions of self-responsibility and God's intervention. It is somehow resolved in the faith of the German Lutheran farmer and, perhaps, of any farmer. But to me it was unknowable. Save a chicken from the winter, and it will die of summer heat. In the contradictory schemes of a contradictory God, the weather will have its own hour, and death will have its.

A meal was to be served in the church basement, an open room with square pillars at intervals and a café-style kitchen. The side walls were paneled and hung with remnants of cloth. Fourteen dining tables, covered with white paper, were laid out with salt and pepper shakers. Prepared by the Ladies' Aid, the food was in hot dishes on a serving table: German potato salad, mashed potatoes, scalloped potatoes, Martin Pfannes's mildly spiced sausage, ham, chicken, sauerkraut. We fell into line. A vase of flowers marked the table for Johanna's sons and her brothers and sisters and their spouses. Diana and I sat at the next table with Reverend Westphal and his wife. They had the mannerisms of the long-married and the indelible look of the ministry, calm strength in their faces, and underneath, the fire.

In 1969, at the point that Reverend Westphal, a young family man in his thirties, moved here from suburban Chicago, he did not guess that the cornerstone of his ministry would be to see this rural congregation into its second century. His alternatives then were to remain in Chicago or transfer to a parish in Des Moines, both positions of more prestige and pay. Trying to explain to me why he chose Beaver over them, he once admitted he had precious few clues. ("I was leaning very strongly toward Des Moines. As a matter of fact, I was scheduled to go there with my oldest son to meet with members of the congregation, and, the night before I was to leave, I woke up suddenly and I went in and turned off my son's alarm clock. I knew I wasn't going to go to Des Moines; I was coming here. But why? I don't know.") God's will? From my father I had heard how Reverend Westphal was a godsend to Zion, putting an end to the uncertainty and disgruntlement that followed the departure, in 1967, of Reverend Reimann, who had guided Zion for sixteen years. The congregation wanted Reverend Reimann's replacement to be in his image, someone conscientious and driven, a minister who fit Zion, whose roots were in the American Farm Belt and in Germany, and not one of the liberal reformers, some of whom, it happened, were about to be excommunicated from the Missouri Synod—"just what they deserve"—in the bitter and chaotic theological divisions over women's rights, ecumenicalism and Biblical literalism. The first minister who tried to take Reverend Reimann's place left after one year, and, for the next nine months, the congregation on many Sundays did not know beforehand who might be in the pulpit—a retired minister, or one on loan from another parish, or perhaps a schoolteacher. In the parsonage, wasps began to nest between window frames. Ground ivy crept across the lawn. Brambles grew. The process that a Missouri Synod congregation uses to obtain a new minister has elements of detective work: interviews, recommendations, some privately done snooping; and gut feeling also counts heavily. That done, a congregation sends out a formal "call" to a prospective minister, after which the congregation's own credentials come into play—its location, finances, growth potential, its status within the synod, its ability to boost a minister's career, and, of course, its people. The final decision is up to the minister. Of the candidates who met Zion's criteria, no one, after prayful consideration, was moved to come to Beaver until Reverend Westphal, absorbed in Chicago, found the call in his mailbox. ("Which was unusual in itself, because it's usually handled by telephone. But there it was, in my mailbox.") On a map he found Beaver and found, twenty-five miles north, Standish, where his seminary roommate happened to have his parish. The Standish minister drove over to Beaver, inspected the parsonage and the brick church, took pictures and sent them to Reverend Westphal. The pictures interested him. They showed buildings with a settled, permanent look, a sense of home, although that word did not strike him until much later. Visiting Beaver, he had a favorable impression of the church officers, and they of him. "He seems like someone who knows how to get his way but who doesn't have to get it to be happy," my father said. Reverend Westphal accepted the call and occupied the parsonage. The lawn was cut, the intruders' nests cleared away, mildewed rooms aired out, the weeds and brambles pulled out. Yet he began almost immediately to wonder if it was a mistake. ("I thought to myself, Lord, I'm not sure I should have come.") Certain parishioners gave the impression that no one could replace Reverend Reimann. Things did not go auspiciously. The incident of the vegetable garden trampled by my father's Holsteins was one thing. And adjusting to the country was not easy for the Westphal children. Beaver is as far from Chicago as Timbuktu. He began to feel—and I know this feeling well—that living in Beaver was living out of his time. The perfect minister for Zion would have been his father, the Reverend John Westphal, who felt at home doing God's work in rural parishes. In fact, why not him? An idea was born. The elder Westphal, stationed in Wisconsin, was nearing the age of retirement and resenting it. If he were to become his son's associate at Zion, might not he and Zion and the Lord all be served? In 1973, the Zion elders agreed to call Reverend Westphal's father, and he readily accepted. He became known as Pastor John and became much beloved. His death in 1983 from a heart attack was an event of congregationwide sorrow. ("The year before he died, we went camping, the whole family did, and he told my brother that these were the best years of his life because he felt so much at home.") In the meantime, Reverend Westphal himself had discovered something about Beaver. ("When I am away from here for a time, I find myself wanting to come back. In my whole life I've never lived anywhere as long as I've lived here. To me this is home now.") I knew this about him from our talks. His own past confusion about Beaver had given him sympathy for mine.

In the church basement, rising from the table, he blessed the funeral meal for us, then sat down. "I'm sorry that it's these circumstances that brought you back here," he said to me. "I guess Grandma never really recovered from being sick at Christmas."

"No," I said, although I had convinced myself in January that she was hale and hearty when Diana and I left. "We almost went to see her the last day we were here. It turned out to be our last chance," I said.

"Who can know ahead of time how things will turn out?" Reverend Westphal said in a reasoned voice.

But it was gnawing on me that during our long Christmas visit I had again avoided seeing her. I found myself confessing and as quickly making up excuses. "It seemed like it snowed every day we were here."

This earned an oblique, raised-eyebrow look from Reverend Westphal, but he let the subject rest.

"Look, there's a familiar face," Mrs Westphal said, waving. Ronald, who was table-hopping, came over to ours.

"Haven't seen you in ages," Reverend Westphal said. "Did you get your buck this year?"

Ronald snickered, snickering at himself.

"You missed again?"

"I didn't have a chance to miss. I didn't see a single buck to shoot at," he said jovially. His lack of fortune in the annual deer hunts—not a single kill in nearly twenty years of trying—was a comic legend in Beaver, always good for a laugh. Ronald had had to learn that deer hunting was one thing he could not take seriously.

The talk turned to horseshoe pitching, another matter altogether. As soon as the weather broke, Ronald and Margo would be outdoors priming for the world tournament. "Good luck," Reverend Westphal said. "Let us know how it goes."

"Yes," Mrs. Westphal said. "Stop by once in a while. The men miss having you around. They tell me the voters' meetings aren't as interesting without you."

"We miss your good arguments," the Reverend said.

"Well, you've still got my dad," Ronald chuckled, self-deprecatingly, as though to distance himself from the family reputation.

"Sure do. No one's ever going to tell him how to make up his mind."

Pressure had been applied recently to get from my father a pledge toward the centennial remodeling project. Members of the centennial committee had been by. The school principal had been by. "We need you to set an example," my father was told, but he was unwilling to go on record for the project. ("I know I made them mad, but I told them right away they were wasting their time.") This, the surviving pioneer fierceness, Reverend Westphal had long since come to appreciate for the good things it said about his congregation.

....

My mother kept her hands and eyes on the eggs she was sorting at the long wooden counter and spoke quickly, as if she had been trying to get the information out but hadn't found the time until now, when we were about to meet with Uncle Martin and Aunt Erna. "The back forty isn't ours, you know." She was concentrating hard on the eggs. "It was still in Grandma's name." The rest of her information was that in 1952, in the exchange of ownership between Johann and Fred, the money had been a little short. The twelve thousand in cash from my father's army savings, my mother's Bay City savings and their share of five years of farm income sufficed only to buy the front eighty. It would have been a simple matter to wait until they had the money to buy the forty also. Instead, because everyone wanted the transaction to be complete, the two men struck a good-hearted, family-style bargain. If my father stayed on the farm over the long haul, faithfully working it, the forty would become his on Johann's death. Not a word was on paper, though, so the forty, under state law, transferred to Johanna when Johann died. Now, under that law, it would be split between Johanna's two heirs, my father and Uncle Martin.

"Grandma didn't leave a will?" I asked, knowing the answer, but needing to hear it, as one might ask the projectionist to bring into focus a heartpounding scene in a movie.

"I told you." A shrug. My mother had no answer as to why, in all the years, this had not been straightened out.

My father was not someone who was impatient about record-keeping. His desk had an assortment of receipts, letters, check stubs, every paper preserved no matter how commonplace. For example: The Bay City doctor who in 1958 removed a wart from Johann's hand had removed one last week from the hand of my mother's brother, Uncle John; my father had kept the original bill and was able to refer Uncle John to the same doctor, still in business at the same address! Another example: Last year when it appeared that the Kalamazoo was past the point of repair—the bottom of the iron grate burned through, and the pancake griddle jerry-rigged by Ronald to replace it burned through also—my father dug into his desk for the bill of sale, dated 1952. The Kalamazoo Stove Company and its retail stores had gone out of business, but, in the Bay City phone book, my father found Harvey Rau, whose name was on the old store receipt. Mr. Rau was operating a repair service out of his garage, which held a cache of Kalamazoo of replacement parts, among them a grate bought by Mr. Rau when the stove company folded. Sheer luck—but my father expected to find such solutions in his desk. It made no sense that there was not, at least, a letter of agreement about the forty.

"I guess we'll have to figure it out now," my mother said calmly, as if that was a complete answer, as if she thought the situation both forgave and punished everyone for the many lost chances. She finished sorting the eggs and stood facing the counter, and it was only then that I noticed the little shooting stars of worry flaring everywhere on her face and noticed the brooding, under-slept eyes, and I knew, as she knew, that the future was more in doubt than ever.

"Time to go," she said and put on her coat. My father had backed the truck out of the garage. Diana and I followed my mother out the door. The moon was full, and the barn, drenched in silver, was like a mountainous pile of ashes.

During the short drive to Uncle Martin and Aunt Erna's house, two miles away on Beaver Road, my father said, "If we'd stuck to that deal, Martin would've lost out completely, and I would've gotten more than my fair share." He meant that Johann, back in 1952, expected to leave a substantial bank account for Uncle Martin and my father, but the account had dwindled and disappeared in the expense of caring for Johanna. Other than the back forty, and the little retirement house on it, no other inheritance was left for the two sons. "The way it is now, I'm glad I didn't buy the forty. It wouldn't have been fair." He seemed to be trying to convince himself.

In one branch of the Kohn family there had been such anger over the division of an estate, in which a prime object was a hunting cabin, that four brothers and a sister—virtuous church-goers with grandchildren—cut themselves off totally from one another. When the sister later died, one of the brothers was not informed until after the funeral. In happier days, before the fighting sucked them dry, they had been a gregarious family with reunions every summer. "Read the Bible. What does it say about the love of money?" my father had said. In more than sixty years, not the love of money nor any other evil had come between my father and Uncle Martin. At their weddings, each had been the other's best man, and they had remained neighbors and the best of friends.

Uncle Martin met us at the door. A wiry man of middle height, he had a thinning slick of hair. His heavy-lidded brow with its lower edge of prominence shaded his eyes. Aunt Erna, fair-skinned, had a touch of urbanity about her, as if she never missed a weekly trip to the beauty salon. She asked if we wanted a soft drink. My father shook his head. "Better wait till we're through." He had an expression of internal conflict, of unhealed injury.

For the next two hours the six of us sat around the kitchen table, double-checking addresses in the phone book, writing them on envelopes, and signing thank-you cards. This was the last of the funeral rituals. Everyone who paid respects at the funeral home was to receive a card, the counterpoint to their signatures in the "in memoriom" book. There was no written record of those who had attended only the church service, but my father and Uncle Martin remembered in what order everyone had filed into church, where everyone sat, and from memory they drew up a complete list.

My father had dark pockets below his eyes, in a tired, pale face. Uncle Martin and Aunt Erna's Florida tans—they had been called to the funeral from their winter home in a trailer park near the Detroit Tiger spring camp—were not credible, any more than colors in hand-tinted photographs. My father put a hand on his head and slowly scratched among whitening hairs. He let out a breath from far inside. This was his way of writing. His eyes were dedicated to finishing the ritual. He wrote short appreciative notes on some cards. He signed. The writing was over.

We moved from the kitchen to soft, print-covered chairs in the living-room. My father began to relax. "Don't look so glum," he said to me. He smiled at the corners of his mouth. He asked Uncle Martin if he had seen any Tiger exhibition games.

"No, but this is going to be their year. They've got the best pitching in the division."

"I wish they didn't have so many crybabies. Soon as one of them has a good season, he either wants a million dollars or he wants to play out his option and sign with the Yankees," my father said.

"I can't figure the Yankees. With all the money Steinbrenner's spent on the Yankees, he should've bought his way into the World Series by now."

"The guys he signs, they're playing to make themselves look good, not necessarily to win. That's the difference. You got too many guys who're just in it for the money. They've all got agents now, and all an agent wants is to get rich."

"That's who makes the country go round, agents and lawyers and lobbyists."

"True, true. There's about one lobbyist per farmer these days," my father said. "I suppose you could say I've even got one, now that I belong to the Michigan Dry Bean Association. The association has a lobbyist in Washington."

"What for?" I asked. The dry-bean farmers were unique in their trust of free enterprise, neither asking for nor receiving price supports from Congress.

"He's just for show. We like to look like we're part of the crowd," my father said, winking. He sipped on his drink. He looked thoroughly relaxed. "No, I'll tell you. He doesn't lobby for subsidies. He's in Washington to keep the foreign markets open, Europe and Japan. The U.S. wants to put a tariff on French wine, and the Europeans say they'll get even and stop buying our beans. He's supposed to prevent that. Sales, that's the whole point of the association. The latest gimmick is this old milk tanker that's been converted into a big mobile bean kettle. It has a steel jacket of hot oil to cook the beans. They drive it to shopping malls and feed everyone free baked beans." Inside him must have been emotions that had nothing to do with this quite ordinary conversation. But he kept talking. "Every time I sell beans, Oscar takes out a nickel a hundredweight that goes to the association. He's got no choice. It's all automatic. They couldn't get farmers to unionize, so they put us in an association. Amounts to the same thing."

"Just like at the Post Office," Uncle Martin said. "Course, I got a pension out of it." Retired now, Martin had been a mail carrier in Midland for nearly forty years despite one arm withered by boyhood polio. Every chance he had for an indoor job he turned down. ("The outdoor life—nothing can beat it!") Summer evenings he was out in his garden, a miniature farm, until the last light. Weekends, he and Aunt Erna went camping up north. So much of Martin's boyhood had been spent indoors, because of the polio. Do-it-yourself therapies with tubs of Epsom salts and a contraption of ropes and pulleys could not bring back the arm. He built up his other one, and with it learned all over again how to hoe and milk cows and play ball. During my father's long wartime absence, Martin had been the Kohn son on the farm, filling in, waiting for my father's return, about which Martin had no doubts. ("Everyone knew Fred would make it back. What was going to kill a stubborn bugger like him?")

Somehow we began to talk about that period, my father off at war, and he said abruptly, "The closest I came to dying was in Italy, up the road from Palermo." The story went like this. He was at a beach with several buddies on a Sunday pass. Either they ignored signs about the tide or they entered at a spot that wasn't posted. The place looked harmless. The water was no different from a bath. My father, a two-tone figure with a fiery tan that stopped at elbows and neck, waded out and lay down on a sandbar. The afternoon ebbed. The tide came in, and seawater poured over his little isle. He lost sight of the shore. His only view was the forcible rush of waves approaching, rising, pounding over the sounds of horseplay around him. Starting back, he went under. He felt lightfooted and insubstantial. A riptide had him. What was worse, he had no idea what to do in water over his head—it was his first time. He crawled along the bottom. Grabbing for air, he swallowed more water, but his total inexperience, as much as anything, probably saved him. He broke the surface and heard someone say, "Take it easy," and he stopped struggling. He let his buddies push him on the waves back to the beach where, a week later, in the same afternoon tide, two luckless GIs were drowned. My father told his story as pure comedy. He was amused by his inability—it was after all very absurd to be surrounded by the Mediterranean when you couldn't swim a stroke—and he pulled us into the joke. We laughed, and the laughing carried convivially in the air.

He sank back into his chair. His voice dropped to a lower, more natural level. He asked Martin and Erna about their plans for the next few weeks, and Martin said they were returning as soon as possible to Florida by plane or rental car to retrieve their own car—they had flown up to Michigan immediately upon my father's phone call, buying one-way tickets, not thinking ahead—and, once down in Florida again, perhaps they would revert to their original plan and stay until May. "Would that be okay? How soon do we have to decide about the probate?"

"There's no rush. We have nine months to pay the inheritance tax."

"You can go ahead and sell Grandma's house. I don't have to be here for that. We can decide about the land when I get back."

"Sounds fine. The sooner the better with the house, I suppose, the way things are with the nuclear plant."

"Probably ought to advertise it in the newspapers. Not enough cars go by there to see a sign on the lawn."

So a decision on the house was made, and the one about the land put off, all in a few minutes. Aunt Erna brought out a scrapbook. The most recent snapshots were of Tom Wirsing, in flowing white robes over a Lutheran minister's collar and suit, at his ordination in Illinois. Tom was a nephew. (Erna and Martin were childless.) Years ago Tom had been in my Sunday School class at Zion when I was a teenage teacher, and when for a while I was puffed up about becoming a minister. Tom had looked at me with round, serious eyes—he was nine or ten—and said he would follow me to the seminary. Never wavering, he had gone on without me and was now the first in the family to wear the collar. "He was the best student I ever had," I said. "He'll be an excellent minister." It was a compliment to Aunt Erna, since he was in the family through her.

"A minister's life isn't easy," my mother said. "Pastor Westphal works seven days a week, and even that's not enough to please some people."

"Same with a writer, right, Howard?" Aunt Erna said. The compliment repaid.

"Yes," I said, "but nobody works as hard as a farmer, right?" Another compliment. Everyone laughed.

"Those days are gone," my father said. "The farmers today are getting as bad as ballplayers. They won't work unless everything is guaranteed in advance."

We went home. Ahead of us down Carter Road a new oil-drilling rig was lit up like a Christmas tree.

....

After the Sunday services, in glinting, clear sun with no wisp of cloud, the congregation milled about. Oscar Ittner spoke quietly to my father: "So you're going to sell your mom's house?"

"That's what we decided. You know anyone who's interested?"

"No, but I'll pass the word around."

The two men walked off to be by themselves, and Diana and I tried to work our way toward Jean. We were asking family and friends for the names of obstetricians, just in case. We had talked about staying a while, a week or two, or maybe—who knew?—considerably longer.

A huddle of friends had formed around Jean, and we waited for her to shake loose. She had on a fashionable pants suit. The old Zion dress code was a lost cause; I counted twenty women in slacks and without hats. But this liberalism was seen as necessity, a fact of life, and did not indicate a general liberalism. The recent sentiment for women's suffrage raised by Roy at Christmas had gone nowhere; the discussion had been tabled. My father had been right. He understood the remnant of Germanic male vanity that was at stake, the idea that men have a higher calling and a special dispensation to do God's will. He understood, without, in this case, believing in it.

Jean came out of her crowd and hugged us. "I'm sorry about your grandma," she said. "Everyone was sorry."

"I was surprised at how many people came to the funeral home," I said.

"Why? Did you think we'd all forgotten her?"

"Yes," I admitted. My calculations about Johanna had been inexact to the last. I had expected that an old lady, removed and uninvolved, would indeed have been forgotten. In a society that honors youth and the youthful, as the news magazines remind us, who would care about her? But that was my failure, the failure of men who study a society through the media and believe that an attitude or trend, once described, is universal. "It's just that she was in that rest home such a long time," I said, trying to explain. "In Washington, people come and go. We're transients. And five minutes after we leave we're forgotten."

"I guess we don't have that many people to remember around here. How long has it been since you moved away? Almost twenty years? And nobody's forgotten you." Jean swept her hair into place. Her look of mischief broke through my embarrassment. "Besides, your grandma was always smiling, and who can forget somebody who's always smiling?" She exaggerated her own smile. All her life Jean had been teased for her turned-up-at-the-corners mouth.

"What happens when you get mad?" I asked.

"Oh, I yell and scream. That helps people remember you, too." She put on a mocking rural drawl, prolonging her words, a pleased comedienne.

"Amen." Her husband, Tom, who had been an usher at the service and had stayed to count and bag the collection, caught up to us and reached out a big hand-grabbing greeting. "Hey, look who's here!" I had not seen much of Tom in the two decades since we graduated high school together. Fitting into the Ittner family scheme of things, he contentedly drove grain trucks for the elevator. The times we did see each other, at the elevator or at the Willard Hilton, the picture he made was of an easygoing, tanned broad-chested man inside brightly colored shirts—like a tourist in a Hawaii ad, all spreading satisfaction.

"Has Sandra found a buyer for her house yet?" he asked.

"A few lookers, no buyers."

"What about your grandma's house? What's your dad asking?"

"Ten thousand. Which sounds like a good deal, but it's pretty small and rustic. Nobody from Dow is going to want it. It'd be perfect for an old hippie, but—

"—they're all working for Dow." Tom grinned appealingly. "How about you and Diana? It'd give you a place to hang your hat."

Once more I was taken aback by the direct suggestion. "A lot has to get sorted out," I mumbled. Around us the feeling of the crowd, of human stirring, had vanished. "We'd better get going," I said.

When we were in the car I realized we had forgotten to ask Jean to recommend a doctor.

....

A young farmer on the phone was in a preliminary stage of panic. Winter was running on; it would be several weeks before his shorthorn yearlings could go into pasture; and the hay in his storage shed was down to three tiers, the bottom of which, protected from bare earth only by black plastic, was molding, turning to compost. He had heard there was hay in my father's barn left after the sale of his cows. "Sure, I'll help him out," my father said to me, "but where was he born? In Michigan we always get winter into April!"

I said that there must have been times when the roles were reversed, when my father had backed himself into a corner and had to hope for someone's good will, and what was striking about his answer—even to me, as much as I knew about him being a tough actor—was how analytical he could be about the commonly held idea that farmers are a cooperative good-neighbor bunch. "I probably remember every dumb mistake I ever made, and you know why? Because nobody came running over to bail me out," he said. "If it's my mistake, it's my responsibility, not Homer Dumont's across the road." This was his old argument, the high wall at the center of his philosophy. I had forgotten how quickly he could make it seem unassailable. On the Kohn farm you were to understand early on that "paying for your mistakes" was not metaphorical but real. Laziness and stupidity had consequences. The farm didn't let you get away with anything. One time I left our stud bull alone in the barnyard—a mistake, even with barbed wire shot through with electric charges, even with a ring in his nose and a five-foot chain dragging on the ring. Wild, conquering chromosomes in this bull were dominant. It took him less than a minute to knock over a fence post and be on the loose. My father followed his trail. He saw more knocked-over posts. He saw corn rows trampled, a lawn pawed up. He made apologies and paid recompense and was glad to get off easy.

Andrew Nutt, a Beaver farmer and Zion parishioner, was gored when his bull busted out. Pinned in the mud, he managed to wiggle a jackknife from a pocket and stab the bull's eye. That eye, bloodied, enraged, was almost the last thing he saw. He was in the hospital for some time, recovering. Paying for mistakes: the philosophy of the just, not the merciful. A cow dog we once had—a beautiful, frolicsome animal, part shepherd, part husky, part wolf—became the leader of a night-running pack, and one night he reverted to the wild, ripping the throats of Homer Dumont's market pigs. He came home with blood thick and musky on his coat. Dale was not yet two, toddling about, defenseless if the wolf part of the dog went for him. The dog himself had a fair idea of his predicament. He crawled on his belly into the darkest corner under the front porch. Coaxing him out, tricking him—a dog I loved!—with my boy's voice, I led him to my father, who was waiting with the loaded twelve-gauge. Paying for mistakes: a philosophy put together from life the way life has always been on a farm. In 1923, with Johanna pregnant, behind on expenses, Johann took a winter job in Willard and wasn't at home when the baby was born prematurely in a downstairs bedroom, a baby christened Walter, christened without delay because he had only five days to live. God was where? In his heaven? In 1934, when Johann's older brother, Herman, was overcome by Depression debts, he took his life. Paying for mistakes: Whatsoever thou shalt sow, that shalt thou also reap. In the 1980s, the U.S. Labor Department reported that farms had become the most deadly of American workplaces, due primarily to accidents involving the bigger, more dangerous machines. Farmers likewise were dying in upward numbers from cancer, attributable, some researchers said, to chemical fertilizers and pesticides. "Woe to those . . ." Isaiah had said. There was also a rise in the suicide rate among farmers in the 1980s, a rate that recalled the Depression, men dying of shame, equating failure with extinction, paying for their mistakes.

My father's view about legislation pending in the Congress that would have forgiven or alleviated some of the greater farm debt was that "farmers don't need Christmas baskets." He quoted John Kennedy: "Ask not what your country can do for you." He was embarrassed by the tractor parades and the farm-auction protests of the American Agriculture Movement, not because farmers were cast in defiant roles but, quite the opposite, because the effect, he thought, was to elict for farmers more public and congressional sympathy and, ultimately, to exhaust their moral capital. In troubled times, the younger, modern farmers, he thought, were looking not to themselves but to the government. Guaranteed loans, price supports, all manner of subsidies and bailouts had made modern farmers dependent and helpless, the fate of the chicken. Modern farmers had incorporated their family names so they could qualify for tax breaks, and so they could, in winter, draw unemployment wages. To be paid for not working! Those years when, in the seasonal layoffs from the sugar-beet factory, my father had been legally free to take unemployment payments, he never had done so. The concept was beyond him. The common view of modern farmers who came into Ittner's elevator was that, as a group, they worked hard and did their best—said at times with a fervor to make Norman Vincent Peale blush. They viewed themselves as the hard-luck victims of government policies not in their power to change, policies that may have been well-intentioned but that went wrong and harmed the very people who were to benefit. My father had his own view, a minority view, which was that many modern farmers, individually, were overreaching and short-sighted and were receiving their just deserts, and that the livelihood of farming, treasured by Americans as a reminder of pride in one's work and of duty to the home, a living demonstration of the "settling of America," proof of cycles and seasons and American square dealing, was being devalued. Farmers, if they were regarded at all by the public at large, found it was now with exasperation rather than reverence—"Let's keep the grain and export the farmers," President Reagan had joked—and they themselves had forgotten self-responsibility, pestering for special attention and putting guilt trips on elected officials. As for journalists, they misinterpreted the whole business, seeing in the beggary and greed of such farmers the disgrace of everyone who ever grew food. This was his view.

I said to him that his tombstone could say NO ONE SHOULD FEEL SORRY FOR ME.

"Well, I've never felt sorry for myself," he said. He climbed into the haymow to throw down thirty bales for the young farmer with the yearlings. "When I go to my grave, I don't want to owe anything to anybody."

"Your army buddies who pulled you out of the ocean, you owe them a little something."

"Nobody lets somebody else drown."

"It was your mistake."

He accepted that. "Yes, it was."

The bales bounced hard one after another on the barn floor, where I piled them, ready to be trucked away. "I count thirty," I called out.

"I'll throw down a few more."

The farmer with the yearlings was a weekender. During the week he worked at Dow Chemical, leaving Saturdays and Sundays for the country life on twenty acres of Beaver land, part of a falling-down homestead he planned to fix up. To date my father was unimpressed—the sheds remained swaybacked, the barn had holes in the roof to let in pigeons by the flock. "He'll putter and putter, but what he ought to do is tear it down." The place had gone downhill while the previous owner, a farmer of my father's generation, was dying from a series of illnesses. I had felt sorry for him. Stopping in once to buy a soup chicken, he stood beside his truck, his mouth open a little, as if in hungry awe at the vitality and orderliness of a working farm. It might have been pride that kept him from asking for a helping hand on his place, although, from what I knew, it was simply that he stopped caring. His buildings caved in while he wasted his time in Bay City saloons. Of him, however, my father was less judgmental. I tried to see my father's perspective: the old farmer, through no fault of his own, was a victim of bad health and its self-pitying effects, while the new owner, with his weekender's pace, it dawned on me, was prolonging the dead man's shame.

More bales bounced down. My father was putting his back into the throws. They were unwieldy and weighed up to sixty pounds. One hit the edge of the mow and popped its twine bindings. It smelled of summer. I kicked the pieces off to the side for the chickens. My father was breathing hard. "You're up to thirty-six," I said.

"There's a lot of timothy in some." The timothy was light and fibrous, hardly worth chewing.

"I doubt he knows the difference between timothy and alfalfa," I said.

"I'm not going to gyp him just because of that." He leaned out from under the heavy crossbeam at the front of the mow as if I should certify his statement.

"He's only paying for thirty."

"I'll throw down four more." The charge was still in him. Four more bales were flung onto the floor. He climbed down from the mow. He brushed himself off. He pulled himself upright. For a moment he looked strong and hard and poetic. The bales were stacked level with our heads, and he went up to the house. I did not feel like following. There was an ironic price to be paid for going it on your own every day of your life, I thought: you never learned whether there was anyone else you could depend on.

The knotted end of a thick hemp rope hung near the bales. I pulled it, and up under the roof a hay sling moved on a steel track. The sling system was original to the barn. We had used the system into the 1950s, when hay was cured in windrows and brought from the fields loose on horse-drawn wagons and was lifted in the slings into the mow; we kept using it even after my father sold the horse and bought a hay baler. My father adapted the system for bales to ride in, although they had to be placed just so inside a sling, a flimsy-looking, floppy thing of thin hemp lines and grooved wood. Wrapped around the bales, it would rise on the opposite end of the thick hemp, pulled by the Massey-Ferguson. From the barn floor the sling went up the equivalent of four stories and hit the central junction of the steel track at the peak of the roof. I used to hold my breath when this happened. Hitting, it had to have momentum to switch onto the track and slide twenty or thirty feet over the mow, and often the hit was hard and jarring and called into question the strength of the thin hemp. Sometimes the full load shook loose and fell. It was a crazy kid's dare to stand dead below the sling, so unprotected it made you feel a blissful high. Pulling on the rope, I remembered the times I'd experienced it.

The rope pulled with no effort. The maple pulleys, rubbed honey-smooth, were as sweet as ever. A few pinpoints of daylight shone through the roof where nails had rusted out, but otherwise the barn was in working condition. I began to climb a wooden ladder nailed to the south side of the barn. I have a height phobia, and my hands began to sweat. The ladder was straight up until the last seven rungs, which were braced out from the wall at a thirty degree angle, creating a triangular space for a hinged ventilation window to open. At this point you had a choice. Stepping inside the space put the ladder safely between you and the mow below. Staying out was to lean with your back unnaturally over air. Blood would begin to leave your feet. From my kid's habit I took the outside route. The ladder finished at the track's southern junction. I reached up to clear away bird nests. Sparrows flew out and played an Alfred Hitchcock scene around my ears. Ducking, I looked down. The overwhelming nature of my height phobia is a desire to jump from high places. The first moment of looking down from the ladder was as panicky as an analogous moment years ago in the Sierras when I almost went over a cliff. I was holding onto the top rung with my left hand. My muscles were shaking, and I could feel a rush. I thought of the *National Geographic* article I'd read as a kid about an African tribal rite in which boys coming into manhood jump from trees, head first, vines tied to their ankles to break their fall at the final instant. Only by jumping can the boys achieve self-respect and status. The article had inspired any number of jumps in this mow, and had, you could say, helped my brothers and I get up our nerve to jump into life—to leave here—although the leaving, like so much in life, perhaps had been imaginary. I kept looking down. The hay on the mow floor seemed a long distance below. I swiveled my feet and kicked them out into the air.

....

Afterward I had a lightheaded feeling of not knowing where, or who, I was. I walked from the barn into a spring thaw. The sentinel maples in front of the house were blurring with tiny, complicated red-and-yellow buds. The network of branches crossed and connected. On the washline my mother's laundry flapped in the breeze as if struggling to learn to fly. I went inside.

The thermometer in the west window of the kitchen was up to sixty-one degrees. The sun's rays that in winter seem to stop a few thousand feet short of Michigan were beginning to reach land. Pungent swirls of ash eddied up from dead logs in the Kalamazoo. My father had let the fire go out. We both smelled faintly, sweetishly of sweat, and we changed to town clothes. My father wanted to go to Auburn and be back before noon. He had a routine: groceries from an IGA store that belonged to the Independent Grocers Association chain; cash transactions to cover his Consumers Power and Michigan Bell bills, handled at the Auburn Mutual Savings and Loan office where teller girls bantered and flirted innocently with him; and a checkup on the blood-pressure machine at the Auburn Pharmacy.

Two cars whizzed past before we could pull onto Carter Road, asphalted like a highway. Its emptiness could no longer be taken for granted. Beyond the asphalt, the ground was weakening. At our first turn, Seidlers Road ran brown and soft. The road was a euphemism for a line of mud. The pickup slithered in tracks that were constantly being reborn. Another truck, driving fast toward us, lost direction in the mud, and my father had to wheel onto the narrow shoulder. "Crazy kid," my father said. "They complain us old guys don't have the reflexes to drive. At least we don't go sliding every which way. That kid, he don't know the first thing about driving in mud."

On my right I could see a long trough of brown water. ''You're real close to the ditch,'' I said.

''I can see it.'' His hand did not waver. ''The last time I got my license renewed they made me take a written test, since I'm over sixty-five. They looked kind of surprised at my score. A hundred percent! 'Course, if they want, they can take your license away regardless of your score. Last year they took away George Hearit's license.''

George Hearit, in his eighties, lived by himself a mile south on Carter Road. ''How's he get around? Does somebody drive him?''

''Not George. He bought himself a three-wheel tractor. You don't need a license for one of those. You'll see him out on the road putt-putt-putting along.'' This was the solidarity of men growing old, reveling in triumphs over bureaucratic prejudice and youthful arrogance.

I kept my eye on the ditch. My father was proving his skill at the edge of the road. After the peculiar excitement in the haymow, a make-believe of danger and escape, my mouth was dry at the nearness and depth of the ditch, lined here with the bare menorah-like branches of elm. The clouds were high, rolling, and mountainous, the light shifting, and in the desert-tone land there were patches of miraculously green winter wheat, so luminous they could have been painted by Rousseau, that went by us on and off to the horizon. At last we came to Garfield Road and turned onto asphalt again.

SIX

On the back forty my father lit a match to the pyre of limbs and branches piled over a dried-out ash stump. Flames writhed and licked but too soon died back. He lit another match, and this time the wood caught. With grub axes he and I dug out brush, heaping the fire. The flames became hot to the point of invisibility, high and white and deceptive. They singed the sleeve of my coat when I got close.

Slash-and-burn is an accepted, timeless method of the farmer. It is a simple one. Advantageously, it adds potash to the soil. It destroys unwanted vegetation with more lasting effect than most herbicides. It does not disrupt, as some herbicides do, the workings of earthworms and microscopic organisms that keep soil loose and crumbly. Slash-and-burn was common among American Indians. The Iroquois, the most resourceful of the northern farming tribes, girdled big trees and burned the undergrowth in piles no different from ours.

From the edges of the widening clear-cut, my father hurled long shafts of brush like javelins into the fire. All the heavy work already had been accomplished. The ash stump was the last major obstacle before plowing. ''Ought to be able to plant navy beans here this spring,'' he said. But his resolve lacked energy and conviction. Today he seemed harassed by recent events, slightly worn, not particularly animated.

''What would it cost to buy out Uncle Martin?'' I asked.

His face went taut. "For his half of the land? Twenty thousand—half the assessed value. I couldn't in good conscience pay less than that." Time and inflation had rendered a double irony. The assessed valuation of the hilly, woodsy forty had increased to forty thousand dollars, ten times what my father might have paid for it in the 1950s. "I didn't buy it years ago, and that's that." There was a finality in his voice.

But I wasn't satisfied. "How much is it worth? On the open market, I mean."

"Hard to say." He spoke slowly. "The forty isn't premium property. Someone like Don can't maneuver his machinery real well, with the woods and the crick and the hill. If you were going to do it right, you'd clear the woods and level the hill, plus lay in drainage pipe and fill the crick in with dirt. But it's not worth the investment, not for forty acres."

"How long would it take you to get your twenty thousand out of it, if you kept farming?"

"Longer than I've got." He put back on his face a distracted weariness that gave him the option to smile. He swung his grub ax with instinctive, casual skill. I listened to the sound of his power, the ax gouging and uprooting. It had given the farm long and faithful service, and it expressed, I thought, a social opinion.

In Nebraska I had seen bulldozers, under contract to Prudential-Bache, Inc., owner of about seven hundred thousand farm acres, push over and clump up a stand of cedar trees; rubber tires were thrown on for extra combustion, and the pile was torched; from the road, the ground seemed to be glowing. Then the bulldozers completed their work, moving earth into an exquisite flatness for center-pivot sprinklers that give the Farm Belt a look from the air of green polka dots. Bulldozers and irrigation systems and mammoth, efficient farm machines dictate that farms expand to hundreds and thousands of acres over which to amortize costs. On a mega-farm, its profile wholly contemporary, no woods for juxtaposition, no reminders of the past, you can only guess at its history. Here, on the forty, I was in the thick of history. The woods, the tree trunks stout and gray, individually dilineated, with round, healed bumps where branches had broken off, and the bristling, pollarded brush, the green shoots of buttercups, the creek banks where muskrats were digging spring dens, this was Michigan as it was before the New World was discovered. It had the illusion of a place still awaiting discovery. A grub ax did not threaten the scene—not like a grunting bulldozer.

....

My father rested his weight on the ax, knitting his fingers around the handle. I could sense a drama in him far removed from the small clearing spread before us. ''I'm going to talk to Martin,'' he said. ''The only thing to do is sell it. Sell the forty.'' He moved away from me, leaving without delicacy in the manner of someone unaccustomed to company. A wind, cool and relieving, stirred the fire. In puffs like smoke signals there rose white, tranquil clouds. He walked between two parallel tracks where his pickup tires had packed the snow last fall, sealing off oxygen, killing the undergrowth, creating a trail to last well into the summer.

I found a stick and drew lines in the mud, a drawing of the farm, of the eighty and the forty. The forty had something special about it, something—as parcels of land go—that disproportionately and metaphysically made it a Kohn place. It was a sort of family metaphor, durable, remarkably unchanged, for all the hacking at it. I could not imagine the Kohn farm without it.

At one point Heinrich had set aside the forty for his oldest son, Karl, who was called Charley, and Charley built a cabin here in the woods. It had one room with a cot and a wood stove. There was no lawn. In summer, among the trees, it had an unknown, secret-keeping look. Charley lived alone here all his adult life. In the 1930 census he was listed as a farmer, a bachelor, and a homeowner, but Johanna, speaking with sad exasperation, better summarized his essential nature. ''He was a lost soul,''she said. She carried hot meals to him down the cow lane and across the stone bridge, and sometimes, yards from his cabin, she heard singing. Whether Charley was in the grip of joy or misery was hard to tell. He poked at the food on his plate with such disinterest one wondered why he bothered at all, and often he did not eat for days. He was sick of himself, he would say—his weepy, boozy ways, his aches and pains, his unhappy lot to be stuck away here. Yet he couldn't bring himself either to leave or buckle down to farming. ("He'd say, 'All you do is work, work, work. To hang with work. Just give me the money.' Soon as he got his hands on some, he'd go off to Willard or Bay City. He didn't have a car—a fortune-teller told him he'd die in a car wreck. He walked a lot, with a sack on his back, and he didn't mind if he slept in a ditch. They knew him at every bar around. The guys would buy him a drink or two, then he'd buy for everybody the rest of the night. All the moon-shiners knew him too. And when he didn't have money he'd sneak into our basement and snitch Pa's wine.") Only a woman as determined as Johanna could have forced nourishment down him. The attendants at a Traverse City institution for the mentally disturbed managed during Charley's last several months, in 1937, to get only grape juice down him. He weighed seventy-six pounds, a woeful thing of bones, stubbornly shaking his head. ("He had been in the crazy house before, and when he got out the first time he tried to turn over a new leaf, but he couldn't change. He was an old dog. So then he gave up and wouldn't eat at all, and there was nothing anyone could do. However long it took, he was going to starve himself to death, and it took a long time. The doctors said they couldn't believe he lasted all those months.")

On impulse I went into the woods to look for the remains of Charley's house that I remembered from when I was young. It was already in ruins then, half a wall, part of a door frame, gap-toothed floorboards, but I could envision the rest, his cot, his jugs and bottles, the strange life. Now, in the fresh growth, unfamiliar and heavy going, I could find nothing. The woods had swallowed it.

Daylight was getting low, the sun in its eternal flame of gases heading for the Great Plains and the Pacific. Our fire had burned down. The ash stump seemed only to be charred. I swung my ax to test it. A fresh flame tore along the ax cut. Fire was set in the stump. In two or three days, unless there was rain, the stump would be a black hole in the ground. An ash stump should be burned. Hardly ever will it rot. Instead, with roots toward bedrock, it will sprout into a scraggly bush. A derivative of brine that Dow Chemical sells by the trade name Dow Flake will kill it, but then nothing else will grow there. There were farms in the valley that showed bald spots where Dow Flake had been drilled into long-gone stumps.

I was spattered with ash. My father was probing the ground with a stick to determine the depth of the thaw. When he finished he threw the stick onto the stump. In the hot ash it did not burn so much as turn to charcoal. His head was down. The work that had been his passion seemed now to lay on him like a burden. I noticed blood on the back of my father's hand, scraped by the sandpapery brush. I was afraid to ask if it hurt. We walked to the pickup truck. The clear-cut was a chaos of bootprints and ax marks, like features of a battlefield, a landscape in which smoke curled from the volcanic, artificial-looking stump jutting into a flecked atmosphere. A crow flew out of the woods, gliding and dipping, its feathers gleaming black, and touched down for an instant on the stump. But only an instant: then it felt the heat, cried once, and beat the air.

....

In the morning Diana and I found the drawing I had made in the mud, preserved in a drying slab of earth. I had drawn the Kohn farm of my boyhood, a tic-tac-toe of lines, each representing a fence, dividing the land into squarish fields. Not so long ago, this was the look of one farm after another. In 1959, fully 97 percent of Michigan farms were fenced and subdivided for a diversity of crops and livestock—fences for wheat, navy beans, corn, oats, soybeans and rye, fences for pastures and hayfields, fences to hold in cows and pigs and chickens—and the average size of a farm was 121.7 acres. Everything that could be said about it could be said, for most of one hundred years, about the Kohn farm. There is a story, perhaps unfounded, that Heinrich's first act on the farm, in an era when non-German farmers generally allowed their cows to roam at will onto land belonging to the railroads (the killing of cows by locomotives caused the first violent populist uprising in Michigan) or onto land belonging to neighbors, was to fence his in: a neat symbol of the man and his Germanic soul, and of the German immigrant assumption that land and religion could be kept apart and through hard work kept purified. Heinrich's fences cut to the horizon in a way that always made me think of the James Dickey poem: "Over hills, through woods / Down roads, to arrive at last / Again where it connects / Coming back from the other side / Of animals, defining their earthly estate"—and the old Germans saw fences more or less the same way, setting off their corner of the earth, recreating peasant Germany. Modern farms have few fences. A fenced, diversified farm in the language of agronomy today is a "limited resource" farm.

"See those stones? That's where our fences used to be," I said to Diana. "We used to pick the stones out of the fields and lay them along the fences."

The stones were visible in rough lines on the winter-killed fields. It was hard not to see lines on this farm, I thought with a pang, and not just because of the old fences. So much of the farm spoke to me of the imposition of boundaries and limits, of stops and false starts, of lines I crossed, ties I had cut. And now? Now my father, who had stood fast, unwavering, unbreaking, the immovable object, who had made the rules, kept the lines straight, held it all together, had begun to talk of cutting loose the one thing that had defined us as a family for a hundred years.

"He must just be talking out loud. He can't really be serious about selling," I said to Diana.

"Selling is probably better than a lot of the alternatives," Diana said.

"He could file a petition with the probate court to get the original agreement upheld."

"A worse alternative, if I ever heard one. Besides, he'd never in a million years do it." We had walked onto the cemetery. Freshly turned dirt was in a mound on the Kohn plot.

"I wish I knew what was best," I said. "For him. For everyone."

"I'm not sure there is any solution that's best for everyone. Not at this stage. You'd have to go back in time and start over."

"I refuse to believe that," I said.

Crows skimmed between the treetops. While not opportunists in all things, crows tend to eat what is available, especially farm crops, but also carrion and garbage, up to a third of their weight every day. In 1972 the crow family was included in the U.S.–Mexico treaty that gave a kind of diplomatic immunity to certain migrating birds. The negotiators from Mexico, a country without regular trash collections, saw more good than bad in crows. The Americans, for their part, added a clause: "A federal permit shall not be required to control crows when [they are] committing or about to commit depredations." Depredations? On a farm, it may be reasonable to define a crow's every act as a depredation. Crows will eat and move on, eat and move on. They can be exploiters on an almost human scale—from Thoreau, they "remind us of our aboriginal nature." The crows overhead seemed to want to remind us of something. Angry, cawing with open throats, they flew directly toward us and distributed themselves in the highest crotches of the tallest trees.

From up there the tombstones may have looked like something once alive flung down from the heights. The crows fell silent. We stopped to gaze up and I noticed—what is easy to forget in the city—how mind-emptying open trees and open sky can be. Your eyes are hit with diffused light, as if molecules are flying free, not unlike a state of consciousness that comes with certain meditations. You are dramatically aware of your surroundings and yet peculiarly remote and detached from them. I could make myself believe that Heinrich in heaven could overhear us.

''There has to be another way,'' I said. ''Selling is the last resort.''

....

Ceremoniously, at the kitchen table, my father laid out the official Bay County map of Beaver Township, Section 21, in which the Kohn farm was outlined by mechanical drawing pencils. With his thumb he pressed upon the map's crisp folds, flattening and smoothing. His movements were deliberate, like those of a magician bringing out his props.

''I don't see why you've got to look at that again,'' my mother said.

''I'm showing it to Howard and Diana.''

''Looking at it isn't going to change a thing.''

My mother needed a diversion. She brought a cigar box into the kitchen and pulled out two envelopes. The return address on one was the Lutheran Layman's League, on the other the Endline Insurance Company (a German name, Americanized from Endlein, one of Zion's founding farm families), but the postmarks were old, and the envelopes made a noise when she shook them—tomato seeds, dried last fall on paper napkins, and Spanish onion seeds, shaken by hand from bulbous flowering heads. She pulled out other envelopes—seeds of cucumbers, peas, sweet william. She arranged them on the table. Leaning, she shadowed the map. I was aware of the scrape of my father's chair being moved, somewhat aggressively, into the light.

''Guess I'll plant the tomatoes and onions,'' she said. In recent years my mother had gone into business for herself, selling baskets of vegetables she grew in her garden. A friend, Leona Kraus, delivered the baskets to a farmer's market in Detroit, where the vegetables, homegrown, handpicked, wiped clean of mud, rubbed to a shine, were hugely popular. ''I'm glad I've got somebody to support me in my old age,'' my father would kid her, although now he said, ''You can't plant tonight. You don't have dirt.''

"I dug some this afternoon. I don't sit in the house just because you're not around." The dirt was in fourteen half-gallon waxed-cardboard milk cartons that my mother had beheaded for indoor starter plots. Two by two, she carried them in from the enclosed porch. She planted the seeds. "Don't put them in so thick," my father said.

"If they come up thick I can thin them. They came up real spotty last year." The smell of earth overpowered the kitchen smells.

"You can't thin them when they're right on top of each other."

"Oh, go on and look at your map." Her actions—the cigar box snapped shut, the milk cartons thumped down on windowsills—were, like his, a little extravagant.

He studied the map, one finger driving his glasses up the bridge of his nose. "The eighty, that's assessed at one hundred and five thousand with the barn and sheds. On paper, that's how much I'm worth." He was making a conscious effort to see the farm according to its legal lines. "I should be worth half a million, if you believe John D. Rockefeller." He looked at us. "I read this article where Rockefeller said if you had ten thousand dollars when you were thirty you ought to be a millionaire when you retire. I had five thousand when I was thirty, so I figure I should be half a millionaire. I could've been, too, if I'd done a few things different. As far as that goes, I could've been a millionaire. I had my chances. Somebody's down and out, that's when you can take advantage. Buy up his land. Buy his equipment. Why should he care? The loan company's going to get it anyway. Plenty of farmers operate like that. No different than Rockefeller. It's not so hard to get to be a millionaire if you don't care how you do it." He rubbed the side of his face in a complicated emotional gesture. I wonder if I should have done a few things different, was what it might have said.

I peered closer to the map, which circumscribed so many years. There was a squiggle for the creek. In Heinrich's time the squiggle ran through a juncture of property lines where the forty met the eighty and where the land of two neighbors, Herman Beck and Peter Schuster, also met. Heinrich had wanted a right-of-way along the creek so his cows could cross unimpeded from the eighty to the forty. In 1904, he paid Herman Beck two dollars for an access strip. Heinrich redrew his map, placing the west side of the creek inside Kohn boundaries. A few years later a feud developed between Heinrich and Herman Beck, who produced an official county map with the right-of-way returned to his province. Had the creek changed course? "The problem was, it was a handshake deal. There was no record of it. Nothing was ever registered with the county," my father said. "Probably it would've cost seventy-five cents or a buck to file it with the register of deeds, and those old guys were tight with a buck." Heinrich was not shy to allege a case of fraud. The feud worsened. Herman Beck dug in a fence to match the lines on his map. Heinrich drove his horse and buggy to see Peter Schuster, the other farmer at the common corner. Schuster had hoped to stay out of the feud. It was impossible. Choosing sides, he took five dollars from Heinrich and gave him access along the east side of the creek. The map on the table showed an eastern strip connecting the eighty to the forty.

"If you sell the forty, what's going to happen to that strip?" I pointed with the tip of a pencil.

"Have to put a fence back up, I guess."

The map, a shiny white, reflected light into my eyes. My father refolded it, a locked door, it seemed, in the wall of fate.

My mother had distributed her milk cartons on windowsills throughout the house. I could see her in the dining room examining a table of potted flowers. She walked back into the kitchen. She seemed extraordinarily pleased. "It's going to be an early spring," she said. "My shamrocks got blossoms on them already."

....

Another young man from Texas had been by with another impressive offer for the oil rights, and, receiving my father's stock answer—"I'm not interested"—he had said, "Boy, I'd love for my boss to meet you. He'd get a real kick. Aren't too many freethinkers like you left." I knew what he meant. On the occasions when my father and mother had tracked me down for a visit, arriving in their pickup, with foam rubber in the pickup bed for naps and a cooler for sandwiches they ate seated on old wooden milk stools that my father had covered with carpet remnants, I would introduce my father to friends as "the last of the great individualists." He had about him, then and now, a deliberate impenetrability, an aura, which I think he was well aware of. He told Diana, when they first met, "I'm an independent," as if that would set her mind at ease, and in a funny way it did. To me it was the status quo; it kept me at the same distance as when I was young and he dominated my life. I could understand the frustration of the lease hounds from Texas who wondered why no amount of money could buy the Kohn oil rights. But I was not bemused. Emerson had written that "a landscape is an armory of powers"—to be taken up in adverse times, as I read it. The lease money was enough to buy out Uncle Martin. "I'm not interested!" What kind of answer was that?

When, at last Sunday's service, another parishioner had chided my father, saying that he was going to miss out on the oil rush, he had replied, "What do I want to get rich for? You know what the Bible says about the rich man entering heaven." He said this with clever, practiced nonchalance, a perfect Germanic statement, revealing nothing, revealing everything. I had tried years ago to figure out my father's attitude about money. First of all, it was the love of money, as we knew from Paul's letter to the Corinthians, that was to be feared. Loving money too much led people into dirty tricks and callous, selfish lives; it made them uncaring and ultimately unhappy. But—and this was harder for me to see—money itself could be as bad. It could induce lapse and sin. My father's theory of the rich was not the populist one that they are cutthroats and bandits but the Germanic one that they are lazy bums, worshipping indulgence, and as able to walk through the gates of heaven as a camel through an "eye of the needle" (the term for a narrow entrance into the city walls of ancient Israel). He himself never ceased to emit the sense that he deserved heaven—and a seat in the German choir, I might add. In one of our exchanges, one of my collegiate runs at him, I asked him, "What if overnight you became a millionaire, without wishing for it, without doing anything wrong to get it, would that be okay?" To which he said, "Once you've got it, it's hard not to love it." I thought he was avoiding the point, but later I realized I might be the one who was being narrow. Perhaps easy, unaccountable money could ruin you as truly as money that you cheated or did harm for. How many big-dollar lottery winners curse the day their number came up? Better to make one's way, as my father and mother had, on small economies and diligent hoarding. Of course, if you played the lottery, didn't that mean you had a secret love of money, that your heart was not pure to start with?

It was another of our old arguments, unresolved and probably unresolvable. But back then I was going my separate way, and what did one argument more or less matter? Now that I was home, however temporarily, it mattered. Why not take the money for the oil rights? He had said at Christmas that he didn't want the trampling intrusion of the drillers or an eyesore hole left behind, or, if a gusher came in, he didn't want to be plagued by the noise of an oil pump, jackknifing in its peculiar rhythm, a Texan rhythm. But beneath this, beneath the possible violations of his land and his peace, was, I believed, the corruptibility of money. His was the fearful self-denying moral judgment of the German peasant, of someone green and unripe in the ways of the world, of someone untouched by life. To be "untouched" at age seventy was—not in a praiseworthy sense—to live in a dreamlike state, to be blinded and paralyzed by impractical independent thinking, to live outside reality. Or so I believed.

In Nebraska I had heard a farmer say, "A guy owes it to

himself and his family to take whatever he can get.'' He was talking about the U.S. Department of Agriculture's PIK program, by which farmers who idled their cornfields were paid a set amount of corn, a program meant to give farmers a year's crop without expense, time, or risk, and also meant to reduce the surplus of cheap corn and jack up the Board of Trade price. To many farmers the availability of PIK corn had the same manic, heaven-sent quality as an oil boom. ''I look on it as God's gift to farmers,'' the Nebraska farmer had said. ''Why the hell shouldn't I take it?'' Why not, indeed? But to my father it was ''farming without working,'' which is to say it wasn't farming. Also, as usual, the farm operators who least needed the free ride received the largest share of PIK payments—millions to the farms of Herdco Inc. and Prudential-Bache—which seemed to prove his point about sloth and the rich.

As for me, what I saw wrong with the PIK program, after the fact, was that it failed in its ambitions. The corn surplus was only marginally reduced; prices improved not at all; and farmers continued to drop out of the marketplace. The program proved to be a come-on and a hoax, and it played with the hopes of those farmers who thought things were going to turn around for them.

I also came to see, after the fact, that a profound wisdom was at the heart of my father's refusal to go near easy money. (''Nothing in life is free.'') In November a farmer had showed off the locations on his farm where petroleum geologists thought there was oil, pointing to dream holes in the ground, to lightning rods of steel not yet in place, to the inhaling sound of pumps not yet in hearing. ''It's like the lottery, except they pay you to play,'' he had said, gesturing hugely, so certain of striking oil that he had bought an expensive motor home in which to tour the country this spring while his neighbors were in their fields planting. The likelihood of a strike had been endorsed by no less an authority than the Sun Oil Drilling Company, which had leased office space in Midland for five petroleum geologists and a support staff of sixty-five. The geologists saw the Saginaw Valley different from its inhabitants; theirs was not a roadside view of an agricultural colony, not a view of farm fields that were a losing proposition, not a view at ground level. They set geophones in the ground and blasted with dynamite, bouncing sound waves off underground formations, and saw places of hidden wealth. On Waldo Road the company had two thousand-barrel storage tanks, set on struts, the underbellies oil-stained from constant fill-ups and discharges. Weekly, tanker trucks heavy with oil rumbled down Waldo Road, breaking chuckholes in the pavement, making runs to a refinery in Saginaw. You could get the impression that once the geologists designated your farm, royalty checks were a foregone conclusion. That impression was wrong. The farmer with the motor home had had his moment of truth a few weeks ago on a cool, blowy day. It had been a long, dragged-out moment—the prayer under his breath, the spit of sugary brown dirt from the hole, the intervals of shutdown to cool the metal and to check the depth and to tighten bolts on the portable rig and to smack recalcitrant ones with a wrench (although the machinery was sophisticated, the drillers still had a rough-and-ready side), and then the final shutdown, the hanging heads beneath the steel skeleton, the farmer kicking dirt, the bleached silence until he spoke: "Hope we get an early spring. I got a lot of plowing to do."

You had to have known him before his period of near-wealth to understand how the oil rush had almost destroyed him.

....

My father's first forkful of chicken manure sailed through the open coop door and into the wooden box of the manure spreader. His fork had nine blunt tines, closely spaced, and had the effect of a shovel. With chicken manure it is not the weight but its loose packing that brings difficulty. "Watch it. You're putting too much on your fork," my father said to me. It was dark under the coop's low, sloped roof, and I stumbled and spilled the fork. Manure blew about. We began to cough. "I told you! Jimminy! Okay, we better take the windows out. Get some air in here!"

From her garden my mother watched uneasily. Cleaning manure from the chicken coop in early spring had about every condition a job can have to trigger a heart attack short of a ghost at the bottom of the pile. It had close quarters, the dust of chicken droppings, a compass of stifling air indoors and chilled air outdoors, repeated pulls on several muscles, and a precedent. My father's first heart attack had come in 1981 while he cleaned out the coop. He had stopped to sit and catch his breath. Cholesterol deposits were gathering in a plug where a secondary vessel enters the heart, a doctor said later, "But there's no permanent damage. Consider it a warning from God." My mother had said that sounded right.

A time of doubt seems to come to many Germanic farm marriages when old men's hearts begin to fail and patterns of support and dependency are upset. There is a different sense of who is taking care of what, or whom. No longer: Please take care of this for me. But: Please take care of yourself—please don't die! Oscar Ittner, working at his elevator, was scheduled for triple bypass surgery. Ted Kaiser, a neighbor, was dead of a heart attack while shoveling snow out of his barn. The morning Ted fell, my father was a mile away in his barn, snow shovel in hand. It was easy to believe the men were too willful for their own good. But did the men have a choice? Retirement was idleness masquerading as freedom. If they just kicked their feet back, they might never get up. What an ironic, bittersweet time for their women, who, somehow better at pacing, more in step with the years, were entering their best ones. The 1970s had been good ones for my mother, an expansive, freed-up period, her children grown but not yet scattered, Johann and Johanna moved to their little house, my father in good health. She had begun to market her vegetables in Detroit. She had joined the quilting bees, her first bit of prominence at Zion. At Sunday worship she appeared in dresses bought with her vegetable money. The ushers began to address her as "Grandma," which pleased and tickled her to the extent of telling me about it in a letter. The discussions my father brought home from the Zion voters' meetings he shared with her, no longer with Johann and Johanna. In the fields my mother worked well with my father, better than us kids, because she understood where he was in his head and what he planned to do next. Years before, she had mooned for a time alone with him, and, as the dream came closer, she had been "on pins and needles, worse than her wedding day," as a friend remarked. She could be heard humming, "You Are My Sunshine." But in a dream begins responsibility. Now, in the 1980s, she wished to God she knew how to keep the dream, and him, alive.

This morning she had said, "The weather's supposed to hold for three, four days." The implication that my father go slow in the coop was not lost on him. He tightened his mouth in lieu of a reply and seemed determined to be done by noon. Already the first load was in the spreader, which was hitched to the gray Ferguson. My father started the tractor and drove off to the back forty, his back firmed into the tractor seat. By the icehouse was a slice of tree trunk, my mother's new chopping block. I sat down. My mother walked over and said, "The doctor says he doesn't have to retire, just slow down, but do you think he knows how to do that? As long as the farm is his to take care of, he's going to keep on doing the same as always."

"So you think he should go ahead and sell the forty?"

"Some days I think he should sell it all. But then I stop and think that will kill him faster."

"Has he talked about selling all of it?"

"He's brought it up. It's something we got to talk about," she said. "We'll have to wait and see, I guess. Things will work themselves out. They always do." She returned to her garden. My mother has the specific farm woman's talent for putting a brave front on all situations, the sturdiness that got the early settlers through lost harvests, scant Christmases and early deaths, that was represented in the folk paintings of Grandma Moses and practically canonized by Laura Ingalls Wilder. But, off by herself, her movements were demoralized. She wandered back and forth between rows of strawberry plants. Then I heard her yelling at onion seedlings set out too soon and browned by a frost. She had only herself to blame, and so, for a moment, was free to think only of herself, was free, I thought, of the fates that had in their grasp the rest of her world.

With the second load of manure, I went along, riding on a tractor fender. The spreader sprang into action at the touch of a lever. Its flailers spalled and powdered the manure and spun it onto the ground. Before the Green Revolution, animal manure was at the front end of a three-year cycle. The first year, most farmers planted an intermingling of grain and alfalfa on a manured field, harvesting the faster-growing grain. The second year, the alfalfa was cut for hay. The third year, in summer, cows, which leave their own benefit, were run on the field, and in the fall the alfalfa, a legume that transfers airborne nitrogen to the soil, was plowed under, and the next spring the cycle began again with more manure.

Additional loads of manure were in the coop. Again and again my father lifted large forkfuls, going past the point when I, sneezing, arm-weary, was certain he had to quit or take a break or lose the fork's balance. He made it seem my help was beside the point. Stink broke from the manure, giving me a headache. My father did not notice. He had lost much of his sense of smell, lost to a bad pneumatic cold. Then, in the middle of a forkful, he stopped and walked out. "Getting hot," he said, and it was. Under my jacket, my flannel shirt was sticky. My father admitted to pain and dizziness at the back of his head. "From those blood-pressure pills," he said. "They make you feel like your head's coming off. Half the time I don't know why I take them." He gulped air. The tendons in his neck were rigid, as if to suggest chest pain. His chin was thrust out. He looked at something in the distance. I was shaken. We rested and talked. The rest and the cool air did away with the worst of his headache, whatever its cause. Back in the coop, my father stuck his fork into the manure, studied the situation, removed his fork, studied some more. Aggressively he stabbed it back into the same general spot—and stabbed and stabbed. A rat jumped out. The rat whipped its tail and tried to find an exit. An elementary knowledge of my father would have accelerated the rat to warp-10. It was once a planned, annual event for my father to beat to death dozens of cornfield rats in a matter of minutes. This was before the corn combine and even before the corn picker, when we used a corn binder to cut the stalks at their base and tie them into bundles that we stood upright into shocks: the immemorial, Rockwellian corn-harvest scene. The shocks dried in the field, and in late October we undid them and loaded them onto wagons, moving around the field in circles, gyroscopically inward, until, on the last afternoon, one shock was left in the center of bare ground. It teemed with rats: a setup. My father had let them escape from one shock to the next, into this single scratching, biting colony. A handheld killing tool is preferred by those who have an idea about the psychology of the rat: cunning and artful, an adapter, an aggressor in packs, a killer of unattended small children, and, beneath it all, a coward. My father's weapon was on iron bar. Primitive, metaphysical, and fearsomely practical, it killed with one stroke. After a slaughter, there seemed to be deterrence and not just metal in the bloodied bar, the sight of which, I swear, could stop a rat cold. But the ugliness of the slaughter, the impermanence of the fear, the inevitable rebuilding of rodent confidence: it was understandable that farmers would enter into the magic of rat poison. ("Which would be okay if that junk worked. The thing is, the rats get so they like to eat it. Same with weeds. Chemicals don't bother them either.") In the thick of poisons, every pest was thriving, every one redesigning in a few generations the genes of immunity, improving on evolution, divorcing from the chemicals their animating principle. ("You don't get less rats. You get more rats, bigger rats, super rats! Same with weeds!") A good cat might do better than rat poison. My father's cat was a big gray hunter. "Here, kitty, kitty!" The cat appeared, a running blur. The rat negotiated the step up out the door, too late. The pounce was final; all that would be left was tail. The killing was tonic. My father and I made new claims on the manure. The strain left his face. Soon, far sooner than was imaginable an hour before, we were done, and, in a sense, it was an occasion for disappointment because there had been no more rats to kill.

In the house, the soil in my mother's indoor garden was dry. She brought up a jug of rainwater from the basement. "I'm almost out," she said to me. This time of year there should have been plenty of rainwater, collected under the eaves in a washtub, but the eaves had gone down in a storm.

"Tomorrow I'll get the ladder and hook the eaves back up," I said.

My father objected. "I can do it." It was Germanic chauvinism raised to a principle—that survival is directly proportional to a man's strength, and work was what strengthened him. Work pushed anxiety to the back of the mind.

....

As my father's son, I found the question of the farm's future a difficult one. As a journalist, it was for me somewhat easier. A journalist has a set approach to difficult questions: a flip through the Rolodex, a calling on the experts, an analysis of their answers, cross-examinations and second opinions, a kind of balancing act, all sides taken into consideration, then a confident decision that can occur like magic, an embrace of the answer, doubt and confusion swept away.

I thought that I had come up with a good answer for the farm, a fair answer, a bargain between my generation and my father's generation—and the answer was to convert the farm into a research facility. New Farming would pay to learn some of the tricks of Old Farming.

A year ago I had been on a magazine assignment in the Red River Valley of Minnesota, a glacial lake that with the Green Revolution became an idyll of mega-farms and where, in the town of Roseau, a sign announces that Bob Bergland, the former U.S. Secretary of Agriculture, is the valley's most distinguished native son. Bergland had been gone from the valley several years, off in Washington, when on a Christmas visit home in

1978 he had a talk with two old neighbors, Paul and Dale Billberg. The Billbergs, father and son, had a farm that was a successful mix of New Farming (fifteen hundred acres, wide-body equipment) and Old Farming (crop rotation, animal manure, essentially no chemicals). At first they had been looked on by the agribusiness establishment with suspicion. The county agent for the U.S. Agriculture Department Extension Service told me, "In the eyes of practically everyone, what they were doing, going 'natural,' was going backwards to the days of lower yields. People here are proud of their production records." But the Billbergs had yields as good as anyone's. (Later Dale Billberg was to be named farmer of the year in his county.) Bergland, as agriculture secretary, was intrigued, and commissioned a study to find like-minded farmers around the country. The Agriculture Department is a world-class research outfit, but the study team started at a point approaching zero. No one at the department had ever tried to catalog those farmers who rely on Old Farming methods. The team had to seek out Robert Rodale, the Pennsylvania visionary and mail-order publisher whose company has made a fortune from magazines and books that tilt at the conventional wisdom about chemicals. Questionnaires sent to Rodale subscribers yielded a list of Old Farmers, and several were then interviewed by Bergland's study team. The team issued a report in 1980, which, in sum, declared that Old Farmers could be competitive, in a variety of circumstances, with New Farmers. The report opened up a potential area of new research. Members of Congress asked the Agriculture Department if any Old Farming methods could be adapted to New Farming, and a plan was drawn up to set up research projects on a select number of Old Farms. I knew one of the congressional staffers—I had an "in"—and I thought, at Christmastime, that I could influence the process so the Kohn farm would be one of those selected. The only problem was that this would take time. The Congress had yet to appropriate the specific funds, and the Agriculture Department, with Bergland replaced in the Reagan Administration, had other priorities. Nevertheless, in the Agriculture Department's regular budget there was about $5 billion a year for general research, much of it dispersed to land-grant colleges. So I made an appointment at Michigan State University in East Lansing to speak to a member of the university's agricultural research staff.

We met in a faculty lounge, sitting on chairs of single pieces of rounded, colored plastic that looked like anchored balloons. Michigan State is the original member of the land-grant college system established by Congress in 1862, when the South, until then dominant in American agriculture, was absent from Capitol Hill, fighting a war and losing its place to the Midwest. The state of Michigan, which commissioned the first Northern troops for the Civil War, received the first grant—land and research subsidies that went to Michigan State to bring farming into the Industrial Age.

The Michigan State researcher spoke to me in a familiar language. The adjective he most often applied to modern farming was "efficient." "Efficient," too, was the processing, the packaging, the transporting, and the supermarketing of the farm harvest. The whole of American agriculture was invariably "record-setting" or "the envy of the rest of the world." Its assets were a fabled $1 trillion. With all the government patronage, he said, American farmers existed in a state of grace. Such a difference from the low-caste peasantry of the Old World! In America, when farmers lose their land, they are victims most often of "inefficiency," he said, or, once in a while, bad luck. "In any economic shakeout, a few innocent people will always suffer," I was told.

I said that my father was in both categories and in neither. He did not have the wide acreage that is the definition of modern efficiency, but on the acres he did farm, his yields were above average. His luck had not been the best, but it was not bad luck that had put him in a predicament. What did luck have to do with growing old?

He and his farm, I said, were paradoxes. He had refused to apply the wonders of the land-grant laboratories, and yet his crops were somehow immune, or so it seemed, to bugs and fungi and other pests. In his own milieu of research—comparing his farm to those of his neighbors—he had observed that Japanese bean beetles and the gluttonous corn borers had less appetite for his crops. He could not be certain why. Perhaps his chemically free crops had an unnatural taste to bugs that are accustomed to herbicides. Perhaps his crops had a healthier constitution. Perhaps by rotating crops he kept the bugs from settling in. Or none of the above. But why not attempt to find out? Conduct an experiment. Put his farm into your research program. Control the variables. Determine for yourself whether bumper crops can be grown without chemicals. This was my idea. It stirred the teenage part of me. It thrilled me with my cleverness.

The Michigan State man shrugged. What would be the point? Ninety-nine percent of the serious American farmers had accepted the terms of the Green Revolution, he reminded me. There was no turning back the clock.

What could I say? My idea was undone by my old argument. It was what I had always said about my father's farm, whose obsolete, unvalued future I had seen so clearly with my teenage eye.

I drove back to the farm, feeling a sudden loss. A heavenly host of stars was out. Fiery gases, dead rocks, disintegrating atoms—they were nothing more.

....

My father had waited up. He had the want ads open on his lap. "Made up my mind. I'm going to sell the whole thing. Except the house—we'll keep the house," he said. His voice was calm and sure, a voice that drove at me and made me respond.

"What are you going to do with yourself?"

He did not treat this as a superfluous issue. "Ma can use some help. There's work for two in the garden. And I'll have time for everything I've been putting off. I've got painting to do, and the well's got to be dug out again. And Sandra's room I want to rewire and redecorate for the times she visits, after she moves away.'' For a moment he genuinely sparkled. His face looked smooth with its spring tan, and somehow younger, as if more life had been poured in.

''You've thought this all through?''

''I've thought it through. Like Ma says, as long as it's my land, it's my worry. And I'm tired of worrying about it.''

''Aren't you always saying that it's when you don't have anything to worry about that's when you really have to start worrying?''

''I've got plenty else to worry about.'' He grinned.

''What made you decide to keep the house?''

''Oh, we might decide to sell it. Ma would like a new house where you don't have to fire up the stove every morning. Lots of new houses around. We could buy Sandra's. Probably get a good deal.'' Another grin. ''But I was born to home here, and, well, I'll have time to fix it up. Already ordered a new gas furnace.''

''And you'll keep the garden?''

''About an acre. That'll grow all we can eat. We'll keep the gray tractor and the tractor shed and the chicken coop and the pigpen and icehouse. A few other things. Everything else will go up for auction. I'd like to get it all settled by December when the inheritance tax is due.''

The way he stressed the date, it struck me that the inheritance tax might be a deciding factor, but he said that was not the case. ''December is my deadline for myself. You know about deadlines—that's the only way some people can get anything done.'' Again the grin.

This was the serenity, I thought, that comes when a choice of such magnitude is made—and that might go, I feared, when there is no longer a choice. The certainty that the land would cease to be his, a certainty that removes responsibility and leaves no more chances to take, could be its own dangerous exaltation. It could release a man from the risk of failure and open up a life of new possibilities, or, at my father's age, a life of no possibilities.

Yet tonight, clearly, my father was at peace. ''It's probably for the best, all things considered,'' I seemed obliged to say.

Upstairs, in bed, I tossed about, too keyed up to sleep. Early in the morning I tried a bath to relax, pouring boiling water from a kettle into the tub to heat the water that came lukewarm out of the tap. The temperature gauge on the water heater was always set to save on gas. Steam hovered over the tub. I edged in slowly, nervously. The water was spa-like. I lay down and breathed deeply. But the soothing water did not soothe me. The water pump in the basement could be heard, hammering, sucking, exhausting the well, and when it shut off I heard my father banging about upstairs in Sandra's old room. I drained the tub and toweled dry and went outdoors. It was well and good for father to speak of working in the garden and in the house, but he was a farmer. A hoe or a sickle or a fork should be in his hand; a green field should lie before him in which to step. His land should be his, yes, to worry about.

I felt the urge to get away. Contradictions surrounded my father's farm, not his contradictions, but mine.

....

The Blackhurst FOR SALE sign was rusting in the lawn, while Sandra and Mike redecorated to make their house a more attractive buy. They had put up new wallpaper, ridding the hall of its dimness and giving David's and Scott's bedrooms bright, sportive designs, and now they had begun on their bedroom. A sheet of wallpaper, delicately patterned, was spread on the floor. Sandra moved a scissors along a pencil marking. She held her fingers at the finish line to keep the paper from tearing. "Our house is going to look so nice we won't want to move," she said, ruefully tossing her head.

"I figured it would've sold by now," I said.

"The real estate agents say it's too far out in the boonies. And it's a buyer's market. Everybody knows they'll have their pick of houses soon as the nuclear plant closes." Sandra put down her scissors and measured the wallpaper against the wall. The busywork with the house made her appear calmer than I assumed she was. She had been the first one to learn of my father's decision to sell. ''Every kind of real estate is cheap right now,'' she said. ''Daddy'll run into the same thing when he puts the farm on the market.''

''I can't believe he'll go through with it,'' I said. ''The forty, maybe, but not the eighty. He'll have nothing left.''

''He says that he doesn't want to do it twice. If he'd going to sell, he might as well get it all over with. That's what he told me.'' Sandra was always the one about whom my father felt easy, who always had a few minutes saved for him—no small matter in her hectic schedule. Last year she had agreed to take on three market cooperatives, overseeing the purchase and distribution of bulk cereals and baking materials, bulk cheeses and general merchandise. In all things she was never less than workmanlike, and she went at everything as if it was work. She was her father's daughter: specific work had no meaning; it was life itself. She and my father always had been close, never wasting time on silly things, never doubting their abilities, and telling each other secrets. Even now that she had struck his heart with the sign on the lawn, they talked as always, and he had said if her house sold tomorrow, she and the boys could have free roam upstairs at the farmhouse, as if it would be the most ordinary event to have his only daughter move home for a month or two and then move away for good.

''He seems harder to figure out than ever,'' I said. ''Some days he seems all worked up inside, and then, like right now, he seems to be taking everything in stride.''

Sandra began lathering paste on the wallpaper. ''I think he's trying to adjust to the way things are and not be so bullheaded.'' She interrupted her paste brush, unbowing her head. ''There's a change that's been coming over him the last year or so, even before all this happened. He's been getting more mellow. Slowly, very slowly, but he's doing it.''

''I see it,'' I said, ''but I find it hard to believe.''

A retired farmer, pleasantly contented? A picturesque grandfather? Can a curmudgeon shed his skin? I had seen bits and pieces of a new man—someone with robust psychological health. But I had also seen an old man with a sleepless face, red, bulging, congested, a silent man, saying nothing for long stretches. My father's silences were famous, and were of two kinds, one making you worry about yourself, the other making you worry about him. The first he could use to devastating effect. On the evening of Sandra's first date with Mike, his 1961 Pontiac Ventura had engine trouble, and when they returned to the farm, about 4 A.M., he backed it into the ditch. The headlights shone up into my father's and mother's bedroom. My father woke and went to the window; realization dawned. He put work clothes over his pajamas, started a tractor, looped a chain around the Pontiac's front axle, snugged the chain into a clevis, and pulled the car from the ditch. Mike stayed in the car, a scared kid. My father said nothing. He made no greeting, and nothing in his manner made one for him. Some dimension of the silence seemed to tell Mike not to bother with a thank-you. He went home. Sandra, awaiting her father's outburst, was at the sink, curling her hair for school. "Better get to bed," my father said. That was the extent of his comments, from start to finish. Sandra was seventeen. She did not expect Mike to ask her for a second date, and when he did, and when it let to Zion's altar, he in naval white, she in lace, she knew those fatherly silences and that long, gruff face would never hold the power over her that they once did. In that sense, Sandra was right: my father had been mellowing ever since he walked her down the aisle. His silences now were mostly directed at himself, his suffering and anger held in, his disappointments held in, vulnerabilities and indecisions and self-doubts held in. It wasn't simply that he didn't let his guard down. Somehow he willed his feelings quiet. But on recent mornings the mask had looked haggard, the opposite of the stoic silence which Germanicism prescribes. The Germanic act of silence implies an ordered confidence. Chaos had come into my father's world. What would he do without his farm? Women's work? Not him! The props of the land, its seasons, its dictates, would be gone. He had said little to me, though, and disliked my questions. Perhaps they were too personal. Perhaps, after so long away, I truly was an outsider.

"Like Mother says, we'll have to wait and see what he does," Sandra said. "I talked to Pastor, and he says not to worry. He doesn't think Daddy is going off the deep end or anything."

"Does Pastor know he plans to sell the whole farm, all the land?"

"Yes. Daddy told him."

"And he thinks it's a good idea?"

"Well, yes. He says it's better to choose your time than to have someone choose it for you. Daddy could stall it for a few years, but sooner or later this is what it's going to come to."

The phone rang, and, hanging up, Sandra hunted through a closet for her softball glove. "All right!" She held it up. A drifting, stirred-up cobweb had collected on it. She wiped it clean. It smelled of the oil she had rubbed in last fall. "That was my coach on the phone. He wants us to start practicing soon as we can. This year we're going to do it! We have to! Most of our girls work for the nuclear plant, so we're bound to be scattered by next year."

"You think you'll make it through this season?"

"I suppose we might not, but we've been talking, and we feel we'll try to stick together through the summer, even after we get laid off. We've been runners-up so many times that we really want to win it all. Just once."

....

Midland is a softball capital. Regularly it is the host city for one of the major tournaments—national, world, fast-pitch, slo-pitch, men's, women's—and more than once a Midland team, qualifying for a tournament, has done the city proud. The Kohn family, too, was crazy for the game. Sandra, as pure an athlete as any of us, was a prominent slo-pitch pitcher in Midland, her name often in the sports pages. She took a highly professional attitude toward the game. Softball in Midland is not a casual pickup game with no one keeping score. It is serious, which explains the ''melting pot'' unity of many Midland teams: players off the Germanic farms in with Midland's chemists and engineers, many from faraway places, but everyone dedicated to winning. In team pictures, in their uniforms, players from the city and the farm are indistinguishable. Only away from the softball fields does one realize that Midland—the Midland of the college-educated—is not a city of the Germanic farmer. There is a futuristic work by eleven Purdue University engineers about an imaginary city named Midland in which life is lived by remote control, without sweat or dirtied hands, and many Saginaw Valley farmers would have you believe it is based on the real city. The Midland where they go in the evenings to play ball or to shop is, during the day, a city of white-coated laboratory scientists and white-collared managers whose work for Dow Chemical is at the forefront of technology and commerce and success—on display everywhere in the city. The Midland County courthouse has fifty-six Tudor-style rooms, with classical muraled walls by New York artists commissioned by Herbert Dow. The murals are made of plastic magnesia stucco, blazing blues and reds and yellows that achieved their final color in a Dow Chemical laboratory. On curving streets, veiled by trees, are homes and churches with flat roofs, sharp angles, and clear, plastic walls, the architecture by Herbert Dow's son, Alden, who studied under Frank Lloyd Wright, and the see-through plastic by Dow chemists.

Twelve miles from the Kohn farm, Midland had almost seemed another world. Once it was a place to which I might escape. In high school, before my period of open rebellion, I had planned to live in Midland and work for Dow Chemical. My senior year I toured one of the Dow laboratories. My guide, a practical man and well-spoken, was more than the chemical engineer he was by profession. He had a philosophical turn of mind. He said, ''I'm in R and D, and I came to Dow originally to discover the secrets of life—what makes the physical world tick. Twenty-seven years later I'm still here and still searching, but I have discovered one thing. Once you move to this town and go to work for this company, you're here for life. Who's going to leave the best place he's ever been?" He suggested I apply for a Dow scholarship. "You'll never regret it," he said. I chased that life for a while, listing chemistry as my major on my admission forms to the University of Michigan, where later I was a student protestor speaking hotly against Dow Chemical's manufacture of napalm for Vietnam. By then I had given up on a career with Dow. What killed the ambition, ahead of the war, was my fear of committing myself to anything for life. But I did not give up my vision of Midland as an ideal, or my vision of myself as a professional. Quite the contrary—I was a writer and editor and investigative reporter, and I had other titles as well: political organizer, world traveler, sportsman. Also, some others, which one would wish did not have to be mentioned: unfaithful son, divorced father, wayward child of God, adulterer, and so on through the Commandments. So many personalities! So many disguises! Somewhere along the way the awe at myself had turned into bewilderment. I could pull from my experiences any pattern I chose, but were any real? Had the real me been lost? How could I have hit so many dead ends after hitting the main chance? What had my father said about the successful life having its own revenge, and why had it taken so long to understand he was right?

The absurdity was that I had made a long circle back to the farm only to find that I had been right, too. The farm as we knew it, the one constant in all our lives, could not justify itself. It had been extended to its outer limits and could be extended no farther. Even my father had faced up to the reality. Indeed, I was the one who now couldn't accept what was obvious to everyone else.

"Right now is a lousy time to sell. There's no market. Land prices are way down," I said, sounding off in public, but not giving away, exactly, all I felt. I had gone to Ittner's, hoping to see Jean, and found she was off on an errand. Her brother, Tom, was at the front counter.

"I always figured your dad would hang on to the bitter end," Tom said.

I ran my hand over a stubble of beard. The last few days, my father's intentions sinking in, had left me feeling restless and bohemian. "The problem is that once he makes up his mind, that's it."

"That's his generation—not ours," Tom said, laughing. Two years ago he had dumbfounded his friends by quitting his job with Dow Corning, a subsidiary of Dow Chemical ("a great job") and coming to work at the elevator. A year younger than me, Tom had gotten his chemical engineering degree in an era when Dow recruiters had to sneak onto campuses through back doors, and he had gone from college directly to Dow, determined to be a lifer. But he had begun to feel at loose ends. ("I got tired of the corporate game, trying to be somebody I wasn't cut out to be, having to grab credit and always watching your back.") One day, taking up his father's standing offer, he had come home. It might have been an immature move, a backward slide from the real world, but it wasn't. The elevator had steadied Tom and given him a new, optimistic mood. He had become the heir apparent. Oscar had promoted him from office worker to elevator manager.

"Ever think about buying the old Ittner homestead and going back to farming?" I asked.

"No. Remember how bad my hay fever used to be, me sneezing like a maniac? Even with the new medications, I don't think I'm up to that. And, well, I'm happy where I'm at. It's been an education, getting to know farming from this side of it." Tom straightened up the brochures on the counter. One said in bold type, THAT'S USING THE OLD BEAN—the new motto of the Michigan Dry Bean Commission. "Hey, have you seen our computer?" He grew excited and sat down in front of a terminal. In no time I had a lesson in the handling of futures contracts, price updates, inventories, shipping, the son doing in half an hour what used to take the father half a day. The computer reminded me that Tom was an engineer, but it also was a sign of his optimism. Like Oscar, Tom had faith that the hard times would pass. Out by the loading docks there were creosoted ties and iron tracks of improbably recent manufacture. How long was it since new railroad track had been laid anywhere in the Saginaw Valley? Wenona, Monitor, Coryell, Rooney, Laredo, Flajole—these were the flag stations, one every three miles, when in 1888 the Michigan Central Railroad completed the line from Bay City to Midland. "Tri-weekly, all winter" was an early motto in white lettering on the red railroad station in Bay City, removed long since by age and weather. The ticket sellers at the passenger windows had been laid off in 1951, and at the freight office no one spoke anymore about moving the country by rail. As of last summer the Midland-Bay City line had one new regular customer, though. Tom had talked Oscar into spending ten thousand dollars on a railroad spur with room for twenty-two freight cars. The cars backed under chutes hooked to L-bolts high in a silo, above which a door cranked open, releasing the contents. Previously Oscar had to move all his grain and beans in long-distance trucks or on lake freighters that went up the St. Lawrence Seaway and past the coasts of Quebec and Nova Scotia, the route by which Champlain opened up trade to Michigan. The difference by rail over one year could be a hundred thousand miles and thousands of dollars in savings. When the spur was laid, the spikes driven flat, the Ittners had thrown a party, an ox roast, attended by everyone around. "A good time was had by all," my father had written in a letter.

Tom said now, "We'll miss seeing your dad come into the elevator. I know especially my dad will."

....

The afternoon *Midland Daily News* carried a report from Joseph Mann, the Midland mayor and a Dow Chemical division manager, that a Fourth of July picnic with free hot dogs, free ice cream and free soda pop was an idea whose time had come again. Once an eagerly anticipated event, the annual Dow picnic had been discontinued after World War II, when the vogue of summer changed to camping and boating and getaways up north. Now it was to be revived and expanded. Everyone was invited, the whole valley, for a kind of thank-you. "We've had rough seas, but the crew has pulled together," the *News* quoted Mayor Mann. All winter the *News* had been full of bad publicity—the federal and state investigations of the dioxin in the Tittabawassee, and the long, ongoing countdown to layoffs at the nuclear plant, which would be blamed, in part, also on Dow, an original partner in the project with Consumers Power. There was an agreement for Dow to buy steam skimmed from the plant's radioactive core, but, after the many missed construction deadlines, Dow officials had declared the deal void, and Consumers officials filed a lawsuit, alleging breach of contract. If people were disillusioned about Dow, however, the disillusionment was not much in evidence. The citizens of Midland had rallied behind Dow with letters to the editor, phone calls to radio talk shows, and support-your-local-company bumper stickers. Dow's director of public relations, Jack Edey, told the *News,* "The people deserve something in return." The newspaper story gave me a new idea. I called Jack Edey, and he arranged for me to see Robert Tennant, Dow's manager of market research and new product development.

At the Midland city limits, I rolled down the car window for a gush of air. It smelled of sulphur. I made a left onto Saginaw Street, drawn to the sulphurous vapors over Dow's headquarters. Below them were smokestacks and holding tanks painted green and yellow and blue-green. The steel supports were red and yellow. The Dow family had long had an obsessive interest in what the eye of Midland beholds. Herbert Dow once paid to keep the city's churches painted. I walked into a spotless, vacuumed lobby, but I was in the wrong place. I reversed my route. Robert Tennant had an office in a low, modern brown-brick building outside the city. He was in his middle forties and was dressed in a tan shirt, open at the throat, and brown plaid slacks. He had barbered hair, and his office had cushioned swivel-back chairs. In

1905, on a visit to California, Herbert Dow had met with Luther Burbank, whose experiments he had followed, and as a result of their discussion Dow Chemical chemists were put to work on lime sulphur and lead arsenate, the first two pesticides the company marketed. It was up to Tennant today to see that Dow trademark products stayed current. He was a contemporary man, and meeting him made me feel suddenly uneasy. History was on my mind.

"It was the Germans who brought agriculture out of the Dark Ages, and it wasn't the German farmers who did it, but German chemists," Herbert Dow once said, and he had a point. The insecticides and herbicides that are thought of as American and as the harbinger of modern farming actually hark back to the nineteenth-century Germans—and even, it is argued, to earlier men. In the beginning God created the heavens and the earth, and in the earth were starter chemicals. Along came man, who created Agent Orange, Roundup, Glean, and AAtrex. It is a matter of debate where one creator stopped and the other one picked up. The flowers of the pyrethrum can kill an insect and, in powdered form, they were spread about on Biblical crops and, much later, were used to extract the specific and complicated insecticides, pyrethrins. Pyrethrum plants—for that matter, all plants, all animals—were once thought to have a "life force" that could not be duplicated in a laboratory, part of a larger assumption that certain things—intelligence, genetic memory, the soul—exist only naturally. So the theory went until, in 1828, the German chemist Friedrich Wöhler mixed salt and ammonium cyanate to form urea, the first organic compound synthesized from inorganic ones. It raised science to a new creative high. The chemical industry, first in Germany, then in America, leaped full-blown from Wöhler's brow. In the 1940s competition between German and American chemists reached the level of war. To find a chemical means to destroy the enemy's food supply, industrial researchers became military researchers. The Dow researcher Wendell Mullison was at the University of Chicago, working under a U.S. Army contract on a group of synthetic hormones, phenoxyacetic acids, that can make a green field lie down and die: the chemical-agricultural equivalent of Little Boy and Fat Man. Peace sometimes outpaces science, and in the case of the laboratory hormones, the war ended before they could be dropped from low-altitude attack planes. The research was not wasted, though. It served as a building block for the Green Revolution, the triumph of which remade the American Farm Belt. The Green Revolution and its monuments were uniquely American. In Nebraska, coming over a hill upon Herdco Inc.'s Thunderbolt farm, I had watched it as one might the blast-off of a space shuttle. The silage retaining walls, six stories high, elevated rapidly. Between Omaha and Denver no rise in the horizontal line is higher; and the fifteen feedlots of shining alloy take up seven and a half miles. The Thunderbolt's provocative immensity symbolizes the Green Revolution, and in parade array across the Farm Belt are more farms with blue Halvastore silos and sun-reflecting buildings, each a smaller copycat of the Thunderbolt, colored metal features on a featureless countryside. My father's farm of gray wooden buildings, existing today pretty much as Heinrich knew it, was a museum piece, and that was my idea. Since the Green Revolution was secure, and so much of the past was torn down, why not preserve one of the Old Farms of the Saginaw Valley for history's sake? And why not give the idea to Dow Chemical in its time of need for good publicity?

Robert Tennant had other ideas. He made not even a token gesture of sentimentality about Old Farming. "It's unrealistic in this day and age, and that's as it should be. Survival of the fittest is how the game is played. Always has been, always will be." Before coming to Midland, he said, he had managed his family's two thousand-acre rice farm in northern California. "You want to talk about family farming, that's my idea of a family farm." I had not met him before, but his attitude was in innummerable brochures and newsletters of the U.S. Department of Agriculture, and it was in magazine articles I had written. I felt foolish. I gave up on my idea.

Robert Tennant lounged in his chair. He grew more personable and put on a big breakfast-table smile. I asked him about the Tennant farm, and he described enchantingly the low, cool rice fields, moated by water channels, and the orchards on high ground. I saw him differently. I saw him with dirt on his hands. Then, on my way out, he regained a measure of corporate firmness: "Don't kid yourself," he said. "The number one reason the old-fashioned family farm went out of business—and it's the same reason ag-chemical companies like Dow are doing such a good business—is that the labor force left the farms and moved to the cities."

....

In the brooder house, a construction of flimsy papier-mâché patched with tin, utterly unlike the other farm buildings in character, my mother tended to a hundred California brown chicks, newly arrived in cardboard boxes. Spring had begun, and it seemed like any other spring. The downy chicks would grow their feathers in the brooder house, and in August the roosters would be butchered and the hens transferred to the wooden coop.

In the tractor shed my father took a heavy smithy hammer from a workbench and swung at the three-point hitch. The crank, which he had absentmindedly stuck behind the tractor seat, was bound up in the hitch. He struck several times, paused, bent over, curled his hand around the hitch's extension arm, tried with muscle to budge it, cursed, hammered again. I was at the shed door. I watched the hammer miss and drop to the dirt floor and watched my father keel over. Behind me I heard my mother, approaching from the brooder house. She took a few steps and came no closer. I turned. She had not faltered or lost composure, but incredible fear was in her eyes. I turned back to my father. He was on one knee. His face was white. The line from nose to mouth trembled. He pulled his lips tight over his teeth. He had no words sharp enough for the pain so it seemed. Then suddenly he was shouting in a voice that was all wrong for the constriction of a heart attack. When he got to his feet and came into the open, I could see his left thumb was bloody and swelling. He shook his bad hand a few times, retrieved the hammer, swung away. The scene uncreated itself.

....

At Zion's annual pancake-sausage breakfast, something curious was immediately apparent. My father, in his chef's outfit, a long white apron and a baseball cap, was enjoying himself. In the Zion kitchen he pronounced a griddle of pancakes fit to eat but ordered a pan of sausages back to the stove. "No, hey, I'll tell you when they're ready." They looked a hearty and reassuring brown. "No, no, they're pink inside."

Running a kitchen was the one trade my father had learned in school—the U.S. Army Bakers' School in Raleigh, North Carolina, Class of 1942. At the Algerian and Italian fronts he had kept the stoves lit and had fired his gun only as a member of the platoon honor guard. In civilian life he forgot about cooking except for this one morning every spring.

"Here, you better let me do that!" he yelled merrily.

Tom Pawley, working alongside, began to rib him. "How many stripes you got in this army?"

"More than you!"

"Okay, boss."

After the parishoners were served, my father sat down to a modest plate of three pancakes and no sausages. "It's a smart cook who doesn't eat his own cooking," Tom said.

"Doctor's orders." My father gestured with dismay. "No more pig fat."

The pig this year had come from Gary Brandt, a onetime member of Ronald's FFA chaper. Parishoners had done the butchering. Money raised by the breakfast, at two dollars and fifty cents a meal, would go to the Parent-Teachers League. Roy said they might buy new playground equipment.

Roy was respected at Zion for the many hours he contributed and for his moderating views, and was someone who could speak across the generations. With Lorie and their children, he had been chosen to kneel at the altar in the Christmas pageant, representing for the congregation the New Family, but he also had a long streak of the German in him. He and the others like him, the fourth and fifth generations of the Beaver pioneers, had a special role in the centennial project. They understood both the history of the church and the desire of newer members to keep up with the times. In a few weeks ground was to be broken and the remodeling begun. At the breakfast tables everyone seemed to be talking about it.

An architect, in sketching out the plan, had added two wings to the church, altering its basic design, which only began to explain the controversy. From the beginning, even as Zion's charter members in the 1880s outgrew services at John Spengler's house and decided to erect a church, they could not agree on the appropriate Lutheran scale of the building. The architecture of a church expresses social and ethnic distinction, and, in Bay City, German Lutheran congregations were building fine steepled Gothic structures of brick or quarried stone. Certain of the Beaver pioneers wanted a church that was classical and ornate in its own right, while certain others, among them Heinrich Kohn, suggested they make do with the heaven-reaching hardwoods that existed on everyone's farm. Heinrich spoke for a general attitude that has since been encapsulated in the word "conservative." Conserving what they had least of—money—was a fairly obvious secret for survival. John Spengler's two children, the first two born in Beaver, died in a fire that destroyed his house and all his belongings, but he had savings enough to start over. This conservatism, after months of fierce back-and-forth among Zion's fifteen founding fathers, won out. "The cost is not to exceed three hundred and fifty dollars," J.P. Ittner wrote in the minutes of the decisive meeting. The church was built of wood on elevated ground that abuts the Kohn farm. The view from the belfry was of Seidlers Road, then a buggy track. After twenty years, when life was more assured, the congregation voted to demolish the wooden church, saving only the belfry bell and the outhouses, and to replace it on the same site with something grand. Over a wood frame went kilned, crisply textured red brick. The steeple was brick and black slate and rose seventy feet. The style might be called Gothic Revival. The cost was twelve thousand dollars, not counting the free labor of parishioners—the women preparing meat-and-potato meals in a cook shanty, the men lifting and hammering, Heinrich working again with brick. The church was dedicated on September 15, 1905. King Solomon once said, in a similar context, "I have surely built thee an house to dwell in a settled place for thee to abide in forever," and, one can assume, the congregation felt as settled. But now the centennial had stirred up old questions of sensibility—between restraint, the Germanic sense of knowing your limitations, and pride, Germanic as well.

....

I had assumed, irrationally, that I could find someone to rescue us—preserve the farm, take the stress off my father and mother, stall for time. But the time was up. "Sooner or later, you have to decide," Diana said in bed. "Either we move back here and become farmers or you have to give up and let things take their course. It's either or. There's no halfway."

"That's a major decision," I said dumbly. "Everything else in our lives would come to a halt."

"I'm pregnant, remember. Our wandering days are about to be over anyway."

"I'm still not going to rush into a decision," I said.

"Howard, you've been rushing away from a decision for years."'

Early in the morning, hazy but not without promise, the sun rising redly like a hot-air balloon, the wind softly working the budded-out maples, we inspected Johann and Johanna's little house on the back forty. It had three rooms, a garage, and the basement where she had fallen. I had helped build it with my grandfather, my father and my brothers, over the summer of

1958. The green roofing and green siding were faded now; white air speckled the green. But it would do for Diana and me until we could make a transition to the big house—and babies could be born anywhere. We could buy new curtains and spade up Johanna's dead garden. We walked into the garage. My father had neatened up the place in simple fashion. The cement floor was swept. We found a stack of "No Hunting, No Trespassing" signs never posted. Johanna's gardening tools were in a corner. A muskrat trap hung from a peg on a roughcut hemlock board. Up the steps, the house was clean and bare, the couch and other furnishings sold for next to nothing last week at Sandra's garage sale; in the last hour of the sale, my father had slashed the prices "so I don't have to haul anything back." The kitchen linoleum was cracked. Lightning had struck the house, exiting through the stopper chain in the bathroom sink, scorching the porcelain. The act of treading across the floors as possible tenants flushed my face. I thought of the first apartment Diana and I had in New York, tiny, in a basement, but half a block from Central Park. We had sublet from a man who was a landlord without owning real estate. He had leased a dozen apartments in excellent locations, and, furnishing them like Holiday Inn rooms, he rerented them at a big markup through the Sunday classifieds for which, in the apartment-scarce city, seekers lined up at newsstands on Saturday nights. He kept his name on the mailboxes so the true landlord did not get wise to the presence of subtenants. Despite the tinge of the black market, he seldom had vacancies. We were glad to get our apartment and felt romantic about its shrunken look, a kitchen stove and a refrigerator stacked in a walk-in closet, thinking it had Lilliputian cleverness. At one point we took a trip, planning to be away a month, but we returned a week early. It was late at night, close to midnight. I took out my key. The knob turned, but the door was bolted from inside. My knocking resounded down the hall. A man in pajamas opened our door; a woman in a robe was behind him. Our landlord had found two sub-subtenants for our month away. I was betrayed. The apartment's romanticism was blown away by greed. I saw what it was. The tiny apartment, inhabited by strangers in nightclothes, was only a tiny apartment; its closet-kitchen was only a closet-kitchen. I walked down the hall, between basement walls that shut out air and concentrated the smells of take-home Chinese food, up the stairs, onto a cold sidewalk, gulping the night air, walking for blocks, stepping over gray snow crusted on garbage bags, realizing that the expensive streets of Manhattan have the look, in places, of a shantytown. I thought of home. Home: the farm. In the distance, across the Hudson, lights flickered like the lights of farm evenings, the lights of fireflies, of families sitting in living rooms, of men milking cows after nightfall, lights bringing to mind the people and places I had left. But these were the lights of industrial New Jersey, of shift workers driving pool cars and cruising teenagers in a late-night rush hour.

"Do you remember the night we found those people in our apartment in New York?" I asked Diana.

"I remember. I had to go by myself and collect our things the next day. You wouldn't set foot in it again."

"I'm sorry," I said. "It was ruined for me."

Why had that night come back to me now? I looked around the little empty house, created in such a burst of optimism, when the grand plan was for Johann and Johanna to live out their lives here and after that, in their turn, my father and mother, and then one of their children. The generations were built into the house: the longest boards, oak and hemlock, had been taken from the double-planked sections of the barn and incorporated here, boards of trees that Chippewas and Sauks lived under, upon which were Heinrich's handprints, and Johann's, my father's, mine. But—how had I not seen this before? —this little empty house was only a little empty house. I felt like a tourist to a place with a high-blown image. I was overwhelmed by its lack of fantasy, the jerry-built look, the ice-cube shapes of the rooms. The house failed to satisfy a meaningful design of history. It was a temporary dwelling or, rather, meant for temporary residents. It was—a bitter pill! —perfect for me.

I was shaking. Diana put her arms around me. Because I had once cried for joy in this house, I could howl and sob, feel the loss, let it go, bury my face on my wife's shoulder. But I did not. I felt moved, but not to tears—I felt moved as a shifting, a heaving-over. Then I straightened up and breathed in, a sharp intake that returned me to my Germanic self.

"Are you all right?"

"Sure."

And I was, until we went into the basement. I saw immediately where Johanna had lain on the cement, patiently expecting my father to find her. I thought of how her freedom had come to an end, her move to the rest home, her slow death there, and how it had put me off and scared me and somehow made me feel sorry for myself, made me wish for the younger Johanna who, when she supervised our chores, used to promise extra pie to the best job—the direct command replaced by a contest, introducing us to incentives: modern ideas and emancipation from the Lutheran dogma that work is its own reward. Had she known she was helping set us free? On some unknowable, unattainable level, had she yearned to be free herself? And if her idea of freedom was life on the farm, how had she kept her good spirits and sanity in that awful institutional room at the rest home? Such happiness was confusing.

The day we loaded Johann and Johanna's belongings in the back of his black Ford pickup, drove down Carter, turned onto Seidlers and pulled into the newly graveled driveway here, she was not with us. She walked over by herself across the creek and through the woods, wanting, I think, to be alone a few minutes. She stood off by herself, too, while we unloaded the truck. My father pushed us to be done before milking time and, setting down the last box, was impatient to leave. I didn't want to go. "Go on," Johanna said. She pulled me aside. "You're getting to be a big boy." I was eleven and close to tears. "It's time for you to act like one. Time for you to grow up." She hunched down to look me in the face. "I'm not leaving you. I'll be right here anytime you want to see me, and when you go off in the world and make something of yourself, you'll come back here someday and tell me about it.'' At the Colonial Rest Home, had her inner smile come from her pride in her family, her pride in me? ''She had all she could want. Not the fleeting treasures like fame and fortune, here today, gone tomorrow, but the real treasures that last forever,'' Reverend Westphal had said, telling us that, at the end, hers was the happiness of living through her loved ones. But in her last years I had not gone to her in my expensive suit, because, I think, she would have seen that her boy had gotten ahead of himself; she would have seen the difference between ''making something of yourself'' and simply ''making it.'' And I had come back too late to make something of the farm. It had been too late a long time ago. I felt lost here, idle and misplaced. ''You're not a farmer,'' Diana had said in the winter. It was true. My ideas were childlike impulses at a time when adult decisions had to be made.

The ineptness of my presence here, the stupid futility of it, produced in me a reeling moment of wanting to be somewhere else, anywhere else. The highways and boulevards of the world, the escape routes of my teenage dreams, pulled on me, urged me into the car. The next day Diana and I were on our way to Washington. I had no plan. I felt as I had as a teenager. I thought ahead only to my freedom. Perhaps the devout could find all they needed within Germanic boundaries. I needed the world, and for the next two months I was swallowed up again by travel. My traveling confirmed who I was.

SEVEN

In June, knowing from phone calls that my father was sticking to his plan to sell the farm after the summer, Diana and I made the trip over the Alleghenies and down again to Michigan. I had fought hard to leave, this last time as much as I did twenty years ago, but I had to go back. The decision about the future of the farm was made—I knew that—an outcome that could not be changed, as we cannot change ourselves. And I had come and gone once too often for anyone to expect me to stay, a feeling that was liberating and oddly frightening. It was this feeling, agitating me, keeping me awake, that was bringing me back. On the road the past two months, the motel beds, always before so adventurous, had given me backaches. At home my comfortable marriage bed felt uncomfortable. My writing was not in sync either; there was a chasm, a void, between achievement and satisfaction. I quarreled with Diana. She had contracted with an obstetrician to deliver our baby in Washington, upsetting me with her pragmatism. And why did I want to drink and get drunk? Not since I was a punk kid had that need been on me. Giving in, anesthetizing my turmoil, only worked terror on my subconscious.

On the Ohio Turnpike, our car pointed toward the farm, I said to Diana, "I'm sorry. I realize your patience is wearing thin."

"I know you have to come to terms with this," she said wearily.

"What you're really thinking is that it's impossible," I said, challenging her neutrality.

"It better not be." She twisted in her seat. "I'm not about to become Mommy *and* Daddy while you keep running around the country. This baby means we both have to settle down, and—" her voice grew more urgent—"you can't settle down if you can't let go of the farm. You have to be able to leave it behind."

Who do you mean? —I might have said. Not me? Not the farmer's son who had turned his back for so long on everything that breathed of dirt and sweat. I had lived with departures, had reveled in change, seen it as a constant flux. How I had fooled myself! Traveling across time and space, I had gone in circles, looping in and out of continuity, going nowhere, staying where I had started. My leavings had failed me.

On many levels, political and personal, physical and emotional—I had to admit I was still a farmer's son. While I had lived all my adult life on the two coasts, it couldn't be coincidence that so many of my assignments took me to the middle of the country. And what of an investigative report I published last year? I had discovered that American grain companies were taking weed seeds and dust and chaff, the very dross removed at a county elevator before a farmer is paid, and were adding this to cargoes of grain shipped overseas. Foreign buyers, charged according to weight, were being shortchanged and were angry. The American exporters had sought an advantage in the international marketplace and instead had undercut themselves with the dirtied grain and, by the way, had damaged the good name of American farmers. The U.S. Department of Agriculture's inspectors had winked at the practice. My report was an important piece of journalism, I felt, and I had first learned of it from my father.

Sandra had said to me once, "You know, you turned out to be more like Daddy than any of the rest of us." I hadn't understood then what she meant, and I was still struggling to understand. So strong in my mind were the differences between my father and myself that I had to work to see the similarities. Yes, we did have the same stubbornness, the same impatience with poor work. My politics, for all my exotic living, had the same basic populist turn for social justice as his. The Magnificat of the Lutheran liturgy had shaped him and had shaped me. "He hath put down the mighty from their seats. He hath scattered the proud in their imagination. He hath filled the hungry with good things, and exalted them of low degree, and the rich he hath sent empty away." I had marched for civil rights and against imperialism and had helped to organize benefit rock concerts. My father, who believed that the truly unfortunate did not live in the Saginaw Valley, gave generously to the humanitarian and missionary work of the Lutheran church. "I don't know how much you know about this," Reverend Westphal had told me at Christmas, "but your dad is the one who has led the fight to keep Zion looking outward instead of inward. He's the one who stands up and talks for World Relief and Missions, and when he sits down there's usually a few seconds of silence, like at the end of my sermons. I'll look around, and I'll see one guy after another smile and nod his head, which means your dad has said it the way they wanted it said. Outside the voters meetings, I know, your dad doesn't talk much about all the good works he's done. He probably hasn't told you." No, he hadn't.

The one thing I did know about my father was that he was a farmer. What was he going to do without his farm and his work? I felt now the same bottomless fear and resentment that I used to have at the Colonial Rest Home, felt my father in Johanna's place, and understood that the two fears were tied together. The same long slide was ahead for my father, upon which he was throwing himself, as I saw it, plunging his life away.

I had phoned him on his birthday last month from New York, in the middle of the day, between meetings. He should have been out planting soybeans, he said, but he was holding off for better weather. He did not sound good. He regretted buying the brooder chicks in April. "What's the use of raising them if I'm going to turn around and sell everything?" He had begun to develop a divided outlook for the farm in which half was momentum for carrying out his own intentions and half was resignation about strangers taking over. "I don't suppose you have any plans to be in Michigan this summer," he said. From his voice, I couldn't tell which answer he was hoping for.

Turning onto Carter Road, I found myself with my old case of bad nerves. Diana tilted forward in her seat with a melancholy sigh.

My father opened the door to the house. It was late, and while I brought our bags in, he went back to bed.

....

The next afternoon, after church services, my father told us of the monster windstorm that had swept out of Canada last week. "We couldn't see across the road, the dust was so thick. It blew a foot deep in the north end of the barn. Took me the better part of two days to shovel it out. Dirt is a lot heavier than snow. And what that wind did to the fields—well, come on, I'll show you."

Diana and I got into the pickup with him. He drove down Carter Road slowly, a Sunday-afternoon drive like the ones we used to take, except the wind had killed much of what we should have seen. It looked as if someone had drawn lines forbidding green growth. My father turned onto Beaver Road. Ahead was a field of corn in which the rows were vibrant. Up close, though, we saw the rows were largely pigweeds and ragweeds, so evenly spaced they might have been planted by machine. This was a mimicry of nature. No cobs would ripen from these bastard rows. The weeds would get bushy and cease to resemble corn. The illusion would be past in a few weeks. Sooner, actually: the field was to be reworked and replanted. Across the road, a member of Zion was on a tractor, Sunday notwithstanding. He was disking under beans that were not going to recover. Everywhere farmers were hurriedly starting the season over. I understood now the headlamps we had seen streaming through the valley as we drove in the night before.

The wind had been cuttingly, destructively out of season. After a freak storm I have known farmers to speculate about *el niño* or Mount St. Helens or a revisiting cycle of history, or, in general frustration, to blame the weather on the federal government. Back in 1972, when a South Dakota flash flood overpowered a Black Hills dam and killed more than two hundred people downstream, I was a reporter on the scene, and I heard farmers, jerking their heads, zipping up body bags, muttering that just prior to the disaster the CIA had been seeding clouds in the hills, which proved to have enough basis in fact to get printed in newspapers and be incorporated into the folk history of the Farm Belt. The old Germans in Beaver, though, more often see through a ruinous storm to Jehovian judgment, quoting Hosea: "For they have sown the wind, and they shall reap the whirlwind: it hath no stalk: the bud shall yield no meal." Hosea apparently was on my father's mind. "We wouldn't be complaining so much about God's windstorm if we'd let more of God's trees stand in the way," he said. The fields with the most damage, new plants sawed to a stalk, had no perimeter trees to blunt the wind. By opening up the land, farmers had made for themselves a difficult farm habitat. Wind, a mundane phenomenon, had been made cosmic.

On the back forty, my father's strategic assortment of trees had afforded some protection. Beans were in his bean field. He had cultivated yesterday and straightened up the field. But the protection had not been absolute, and there were windburned spots. He was left with a choice most of his neighbors did not have. Their fields had to be replanted; his was in a halfway condition. "I don't know," he said. "It might be better if I disk over."

"What would you lose if you did that?" Diana asked.

"Time," I said authoritatively. "You replant this late, the beans might not get ripe before the first hard frost."

My father smiled at my answer. "That's right if you're planting Sanilacs, which are eighty-five-day beans. But I could get new hybrids at Ittner's that are seventy-two day beans. They'd probably get ripe on time."

"But they'd yield less," I said.

"A little less. But I'm not going to get much yield out of here the way it is." My father was escorting us into the field. The creek was to our left. He picked up a stone big as his fist and hit water with it. The water was about three feet deep, well down from spring melt. The elms on its banks were leafed out, and the leaves were dusty. Summer was closing in fast. "I don't know if I have the time to replant even if I wanted to,"my father said. He mulled over his dilemma, second-guessing himself. "Second-guessing" was the word he used. The indecision was not like him, but he made it kind of amusing. On the way home he pointed to Ed Reichard's eighty, in cropshares to Don. The fields were raw earth, but not because of the wind. Behind in his schedule, Don had not gotten around to planting a crop. This year, being late was a piece of luck. "Well, the first shall be last, and the last shall be first," my father said. He was uncontemptuous of Don. He was almost envious. That was not like him either.

Then, once more, he brought up the question of his beans on the forty. "What would you do if it was up to you?" he asked me. It was an affecting moment, no longer amusing. Once omniscient on his land, he seemed to have lost the last of his control, even before he sold it.

"I'd stick with what you got," I offered. This came out like condescension, and I felt I couldn't leave it at that. "The beans that survived should grow to fill some of the places that are burned out. They'll bush out more than usual and have more pods."

"Could be," he said.

....

Dawn was misty and gray. By seven o'clock my father and I were in the bean field with our hoes. "After the nuclear war, weeds will be the first thing to grow," he said. "Try as you might, you can't get ahead of weeds." Unmentioned, although implicit, was his less polite commentary that pesticides often cover up for laziness and ineptitude. On our drive yesterday he had said, "See there, that hill of quack grass should've been summer-fallowed." And, on the next mile, "He should've cultivated last week with weeds that big.'' A farmer who used pesticides could compensate for such mistakes; it was a form of cheating. I remembered one of his letters in which he had taken care to record the news about marriages and hospitalizations and farm prices, and then, departing from this chatty account, had written, ''In the *Successful Farmer* they say I can plant a new hybrid corn, spray on some weed-killer and go fishing for three months, and it'll be ready to pick when I get back. I guess if that's all it takes to be a farmer, Grandma can do it from her wheelchair.'' It was such an emotional letter. It spoke with such pride, and with such a passionate baring of soul, about his unromantic work.

''You have to hand it to Dow, though,'' my father said now. ''The way I hear it, everybody and their brother is going to the picnic to get free hot dogs and ice cream and see the fireworks. It doesn't take much to get people to forget their troubles, I guess.'' I couldn't distinguish sarcasm, if there was any.

''You should go, too,'' I said.

''Nah. Ma and I can beat the traffic and watch the fireworks sitting on the front lawn. But you and Diana should go. Catch a ride with Sandra and her gang.''

''Maybe we will.'' I had not been to a Fourth of July picnic since the ones in the Zion churchyard I had known as a boy, exciting, carnival-like affairs with games and beer that attracted gate-crashers from around the valley. Too many drunken incidents, resulting in higher insurance premiums, had forced the congregation to discontinue the picnics. The one attempt at revival, a picnic in 1976 for the American bicentennial—without alcohol—had been less than enthusiastically received. Gearing up for a second revival, a fund-raiser for the church's centennial project, the Zion voters were undecided about applying for a beer license. ''It's hard to have a German picnic without German refreshments,'' Reverend Westphal had said. Beer also was of concern to organizers of the Dow picnic, who worried that not having any—Herbert Dow's teetotaling had rubbed off on greater Midland—might keep away valley outlanders when the whole point was to draw a huge crowd and promote the picnic as a referendum of support for the company. Then again, some picnickers might bring surreptitious coolers of beer, putting the patrolling police in an awkward spot and risking unseemly headlines.

This had always been an unresolved part of human relations in the valley. The German settlers, for all their obsession with hard work, were also the most likely to tie one on and cause trouble. My own drinking troubles were as a teenager, and once when I was arrested for drunken brawling at the Band Canyon, my father told the Bay County sheriff, "Let him stay in jail and learn his lesson." It took another arrest, the wrecking of a car, several near misses on the road, and, finally, formal leavetaking of the farm and the early responsibility of a wife and child before I learned it. My right to drink was for me the same as my right to be free, and my father may have understood this before I did. One day, while I was bingeing, he had sized me up and said, "You don't have to drink yourself to death like Uncle Charley. If you want to get out of here, go ahead. I'm not stopping you." And I went. After all these years, I still assumed that my father, with his sober abstentions, his dislike of Bacchus, had had no experience with cutting loose, in any sense. His life had been, after all, the Depression, a World War, and the serious, morally weighted farm. So I was surprised, as we hoed and talked about youthful misbehavior, to hear him say, "I had some pretty wild times myself before I settled down. I remember one time, oh, I must've been seventeen or eighteen. I drank too much at a wedding, way too much, and the next morning, oh, boy! I went to sit on my milking stool and fell off. It ended up that Martin had to milk my cows. But I had to go to church—it was Sunday—and I'll tell you, that was the longest service of my life." Something very similar had happened to me: an unforgettable Sunday morning with the entire family in the front pew, Reverend Reimann bearing down, devoting special attention to us because it was Johann and Johanna's golden anniversary, and I, unable to meet his eyes, the dry heaves ripping inside me over and over. Although Johanna tried hard to overlook my state, it was a blemish on her wonderful landmark day. Now, if I had my father read right, he was recalling that incident indirectly, to put it behind us. He seemed to be saying that we both had been boys and were both now men, or something like that. "Well, God must've wanted us to live through our foolishness," he said, and I nodded.

A little later I asked, "Have you decided how you're going to sell the farm? Are you going to advertise it?"

"No. It's going to be by invitation only. I'll ask for bids, and the best bid will get it." He ticked off a short list of invitees, Earl Gerstacker, Vern Chapman, Ross Koch, all of them neighbors or members of the extended family. At first Don was not mentioned, although at one point I thought my father was referring to him. He said, "I'm not going to sell to somebody who can't take care of the acres he's already got, somebody who's out freezing his buttons off trying to combine corn a week before Christmas." There was a delay, like the delay built into radio broadcasts to screen out bad words, and then he did bring up Don's name. "I told Don I'd let him know when I retired," he said. "It's only fair to have him put in a bid, too."

In the woods I heard a whitetail doe, had a glimpse of her through the undergrowth, and abruptly she was in the open. Twin fawns followed, leggy and spotted white, with quivering rabbity ears. I am one of those who is rhapsodic about deer that appear outside suburban windows, but I realize also that the deer are simply adjusting to new opportunities. In Michigan, whitetail deer, once herded north by lumberjacks and farmers, are asserting territorial rights ever southward, in sight of fire-breathing, aluminum-sided structures. Near Detroit, an eight-point buck nudged open a patio door and, tap dancing on linoleum, put a woman into hysterics. Imagine also Ronald's sense of the wild: He may see deer while going to his mailbox, but in the unpeopled north where he goes to hunt, where Johann hunted successfully, the deer are ridiculously scarce. They are south. ("Why not?" Ronald shrugs. "The cornfields are south.") The selling off of dairy cows and the substitution of corn or brush for pastureland gave Michigan in the 1980s an unprecedented, incongruous, unbalanced population explosion of deer.

The doe and her twins cut across the field and descended into the creek. There was no sense of infiltration in their movements. They drank without detectable nervousness. They were practically flippant. "You don't have enough beans to share with them this year," I said.

"Don't worry. They don't eat that much." This was said in a protective tone, then modified. "Can't do anything about them anyway. Once you take down the fences, you're going to have deer."

Eight rows of beans were left to hoe, and the sun had come out of the mist. We took our rest in the woods in front of us, its interior screened off by a bristling stretch of Queen Anne's lace and wild chicory and silver-gray grass, worn with deer trails and singed in round spots by old stump fires. We found a soft spot under a tree. I gazed out from the woods, to its rude, rampant fringe. I had never seen it so wild. Its tall and imposing size gave the area the feel of inactivity, of work not done and civilization retreating. Here was the sad fate of Fred Kohn, capitulating at last to his age, slacking off, letting go of his sense of order—so it might have seemed to a stranger, so it seemed to me. I asked my father if he wanted me to cut the weedy growth with the hay mower. His tanned face, crow's feet filled with dust, split into a grin. "I'm going to take care of it in a day or two. I'll get back here with the scythe," he said. He was again a man glowing with vital power. Moving back through the trees to the bean field, in and out of open spaces, the sun followed him in a golden spotlight. His stride made him seem young. The difference between his self-doubt of yesterday and his confidence of today was the difference between two men.

....

The next day took us out of balmy and equivocal temperatures into an eruption of heat. The atmosphere that prevailed on the farm throughout the day was not relaxed. My father was irritated because he forgot to take a thermos of water with him while cultivating on the back forty. My mother was anxious to finish picking her peas before they overripened. Diana was near dehydration, and for several hours I was hung up talking on the phone to an editor. After supper I saw five large pails of pea pods to be shelled. You had to think: How could two people eat so many peas? But these were for the freezer and might be donated later to church functions. Energetically my mother squeezed a pail between bare knees, a cow-milking grip. My father took his cue from her, and there was a chair and a pail for me. I was dismayed because I had hoped to talk to Sandra that evening.

"Go ahead," my mother said. She did not look up. Her hair was in curlers, and her exposed neck had deep, running furrows and parallel secondary lines. Her arms, oddly smooth, were busy.

I hesitated. "Go on," she said. "Shoo, shoo."

Sandra had a softball game at a Midland park, and, because Diana and I got lost on the wrong side of the Tittabawassee, warm-ups already were underway when we pulled up. Limbering stiff muscles, lobbing balls, the Blackhurst players could scarcely concentrate on softball. Last Friday, at midday, many of the construction workers at the nuclear plant had been laid off, and in the locker room a rumor caught fire that paychecks were drawn on an insolvent account. Minutes later the Poseyville branch of the Midland Chemical Bank had customers backed up a quarter of a mile at the drive-through windows. Cash on hand was soon tapped out (even though the construction company's account had plenty of reserve), and, one on the heels of another, nuclear workers left their cars to demand action. A relief strongbox, arriving ahead of riot-control police, calmed the situation for the time being. But negotiations about the nuclear plant this week had reached the governor's office. The status was day-to-day. Among Sandra and her teammates, no one knew how long their own solidarity, their commitment to finish the season together, could last. Tonight the Blackhurst shortstop, a rosy, elastic-armed woman, had announced she was moving next week to Ann Arbor, one hundred miles away, but would try to commute to the games in Midland. The players, chattering, took this in. "Hey, who cares? Anybody can play short!" "You tell her, babe!" "Yeah!" A ball left the shortstop's hand at big-league speed. "You guys don't watch it, I'll let you blow it by yourselves. I know you'll do it, too. You'll blow it to Select." For the past three years, Select Stationeries, a younger, faster team, had been the nemesis and the superior of Blackhurst Realty. This year, deep into season, Blackhurst had two more victories than Select, a margin that tempted overconfidence. In order to blow it, Blackhurst would have to lose once to Select, lose another game to a lesser opponent, and then lose a playoff to Select. But last year it had happened just that way.

Tonight's game, against a bottom team, was supposed to be a sure thing for Blackhurst. Fans of the opposing team looked on with an aloof fervor as if unsure why, in the assaulting heat, their women did not walk off and forfeit. Sandra, temperamentally and statistically a commanding pitcher, had a good first inning. She concentrated on her follow-through, a crouch, feet split, glove up. A month ago, in practice, with the glove careless at her side, a ball had fired off the bat and struck her thigh, knocking her down and raising an elongated bruise. In the next game, not wanting to become, in her coach's eyes, a "crip," too banged up to be counted on, she had a particular desire to prove herself and had pitched a shutout—a one-hitter!—not to mention that, in her last turn at bat, hitting a drive inside the rightfield line, she had run like a youngster on her bad leg, sliding aggressively into third base for a triple. The ballpark had become the scene of what in the career of an aging amateur ballplayer is the equivalent of being born again. ("Did you ever have a game where you couldn't do anything wrong. You don't feel it's luck. You feel it's the most natural thing in the world.") If only the feeling had an afterlife: tonight, starting in the second inning balls scooted when they should have bounced, infielders were one step out of position, two outfielders were confused by each other's feints and let a ball drop, and Sandra grooved a pitch, its orbit flat, on which a batter's eyes locked and which went for a two-run homer. Sandra lost her fresh, fast-kill look, and, as if by a jinx, Blackhurst was defeated. "Oh, what's the use? We'll never win the championship," she moaned to her disconsolate teammates.

Across the way, on another diamond, Select had two innings to go for an easy victory. The Blackhurst players packed up their vinyl bags and trudged over to watch the two innings. "We can't get enough of a bad thing," Sandra said to me. "Are you coming?"

"Sure." But I was delayed by Tom Ittner, who came walking up. A team on which his wife and one of his sisters played, and which he helped coach, was about to play on the diamond vacated by Blackhurst. "I heard you were on the road again," he said. "What brings you back here?"

"I'm not exactly sure." I laughed uneasily and changed the subject to softball. Tom belonged to the Redcoats, a group of volunteers who organize Midland's world-famous softball tournaments. Teams from Brazil, Mexico, Botswana, Argentina, and twelve other countries were due in next week for the men's world fast-pitch championships, and it was up to Tom to provide hospitality for the Mexican players. "Guess I'll take them to the Dow picnic," he said.

Tom's players were summoning him to hit fungoes. He selected a bat and shouldered it, turning to say good-bye. "Forgot to ask, How's your dad?"

"He was a little up in the air about his beans. Wasn't sure whether to work them over."

"That big northerner put everybody in a fix, especially the guys with big acreage. This late in the season, they're not going to be able to rework everything. They'll have to give up on some of their fields."

"You know my dad, he hates to give up on anything," I said, and heard a lurch in my voice as I said it. "See you around." I waved and ran to catch up with Sandra.

The game involving Select was in the final inning. Select's

mediocre opponent had shortened its deficit to three runs and had loaded the bases. The Select coach stepped out and signaled for a change of pitchers. The new pitcher set off in Sandra a gleeful derision. "She thinks she's God's gift, and since she was on the All-Star team last year, it's worse. Her head is as big as a pumpkin." Entering the game under testing circumstances, the pitcher drew jeers from the suddenly attentive Blackhurst players.

In sports, perhaps more than in life, ups and downs tend to level out. As we watched, the batter bounced a grounder back to the All-Star pitcher, who barehanded it, spun around, saw runners moving to all the bases, and, double-pumping, threw the ball errantly into the outfield. Three runs scored. The batter went to third, and on the next pitch, a solid hit, she scored, denying Select the victory. The pitcher walked to the sidelines without company. "Oh, she deserves it! She deserves it!" Sandra shrieked. "Isn't it awful to feel this good?!"

I was delighted for her and hugged and kissed her. The Blackhurst players in their lacquer-bright uniforms held up beer cans and cheered. Their hope of seizing the championship from Select was recharged. The shortstop, shunned after Blackhurst's loss, was promising again to commute to the remaining games from Ann Arbor and was clapped on the back: "We hear you, babe!"

Afterward, at a pizzeria, I had my chance to talk to Sandra about our father. "Is he going to be okay?"

"Well, his blood pressure is way down. It's back to normal. The doctor thinks it's because of the pills, but I think it's because he decided to sell. In his mind, things are settled. I don't mean that he wants to sell. He never will, you know that. And it's never going to be one hundred percent all right. But I think deep down he feels it's for the best. Somehow or other he's come to feel that. Because otherwise, you're right: It'd be the death of him."

....

In the morning, while my father worked on the forty with the scythe, I hung the eaves. Two of the guide wires had snapped, and I had to cut and shape new ones from coat hangers. After fitting the curved ends of the eaves together, after looping the wires, after nailing them in place, after shaking the setup for a wind test, the eaves tumbled out of their wires to the ground. I hung them again. They held.

"Just in time," my mother said. A thunderstorm was upon us. My mother and I went inside the house. On the back forty, we assumed, my father was hustling into his pickup.

"He better stay put and not try to drive home. Look at that rain!" my mother said. She began to prepare the noon meal. A hot, bright aura was on her, making me wish my father was not cut off from us. "I don't know why he had to go out there in the first place, why he didn't get you to do it. That scythe is heavy. You don't even remember how heavy it is, I bet." She thrust open the cupboard doors to take out the everyday china. Her back was to me. She lowered her voice. "I don't know what gets into me sometimes."

I looked away. There was a wooden signboard on the wall behind the stove that read, "WHEN I WORK I WORK HARD, WHEN I PLAY I PLAY HARD, WHEN I SIT I FALL ASLEEP." My father had posted it as a tease years ago when it referred to his children. Outside, the storm ripped a dead limb off the big maple. It fell with a splintering noise.

The table was set. The meal simmered in pots on the stove. My mother stood at a window, straining to see. She said, "At least, with you here, he won't be out there worrying about me. I know how he is. He starts to worry, and the next thing he's trying to drive home in the middle of this."

The rain pounding on the roof became a clatter of hail. "There isn't a tornado out there, is there?" Diana asked.

"It's too late in the season for tornadoes," I said, but I wasn't sure.

The hail abated, and rain ran from the eaves into the galvanized tub, overflowing the tub. Rain is the final phase of ordinary thunderstorms. It puts a damper on rising hot air and suffocates a storm. But a tornado keeps its distance from rain, following it. On a road it is possible to drive out of a rainstorm and into the full fury of a tornado, and the velocity of tornadic winds can turn the driver of a car into an involuntary Evel Knievel. The first sound you hear from a tornado is nothing, an obliterating silence. There is a moment in Roman history, from an afternoon at the theater in Antioch, 241 A.D., described by the historian Ammianus Marcellinus, when an actor departs from his lines and shouts, "Am I dreaming? Or are those Persians?" Gazing up, the audience sees on the highest balconies the silent Persian army, their bows drawn, the "sudden unknown destroyer." A tornado has the same heartstopping quality. One minute on a September afternoon in 1960, during navy-bean harvest, the horizon was blue-and-white; the next it was black, and in a huge, perfect hush. The shape of a tornado came into view, a dense polluted airborne fluid. Quickly, Ronald and Harvey and I helped my father tie a tarp over the combine. From the other end of the field, where they had been forking beans, Johann and Johanna came running. It occured to them that the creek, in the opposite direction, was a shorter run and a better cover, but Johann had the truck keys in his pocket. The old black Ford was going to get us home. An avalanche of wind came hissing down, and, with a cry, Johanna twisted an ankle. Johann wanted to pull her along, but with girlish animation, she insisted on hopping to the truck, and made it there ahead of my father. He had stopped. Wind was under the tautened-down tarp, curving it up like the top of a balloon. He had to go back. We watched him struggle with the tarp, a sea captain trying to trim his sails, but a befuddled captain out of his element. If more wind were to fill the tarp, the maximum pressure on the twine stays would soon be exceeded—and sure enough, the twine began to break. In the truck cab, Johanna rolled down the windows and leaned out to yell at him. Hairpins dragged from her head. My father made no response. His concern began and ended with the tarp. And then we saw the whole of the sky spin into the dried bean plants. Strands of plant flew up like bursts of gunfire, twisting as they flew, coiling into the barbed-wire fence so adhesively that they were there for a year after. The crop was ruined, but I couldn't care. I felt removed from time; time presupposed a future. I folded my hands and prayed.

The issuing of warnings, started by the National Weather Service in the early 1950s, helps today to limit tornado deaths and injuries to people who are out of communication or are willful risk takers. Farmers probably fall into both categories more often than anyone. In the movie *Country,* it served director Richard Pearce's artistic purpose to have an Iowa farm family protect a truckload of harvested corn from a tornado. The family's peril—including the threat of financial disaster, always important to an American audience—charged up the scene. If our bean field had been reduced to the size of a living room and the six of us to inch-tall miniatures, a tornado could be approximated with a fan and a vacuum hose. Something like that, handled by special-effects professionals in wet suits, was what the *Country* cameras registered and enlarged and evoked into a nightmare, a Hollywood rendering, several degrees of flourish. Perhaps no documentary-film footage could have kept the excitement as high and done justice to the natural violence inside a tornado, but perhaps no movie can deliver up the fear at the center of the attack—not a custodial fear for your property, but a terror in your belly. Yet, fantastically, my father was out in the weather, fighting to reconnect the tarp and keep the combine hopper dry so the beans inside would not lose their margin to the elevator "pick." (And afterward he raised comment in Auburn when he brought in this load snatched from the tornado.) The sound coming at us was a powerful hum made by the torrents of rotating air, rotating to the left in response to the earth's axis. The torrents were white as they plunged with black streaks along each plunge point. The wind keened like a banshee. Dirt got into my eyes. I lost sight of my father. The truck began to shake and rock with a rhythm to terrify the mind's eye. Behind the cyclonic air, in the spent force of its jet stream, lay flattened trees and buildings. The front of the tornado was over us. I saw my father holding down the tarp, and saw it blow loose. Along its trailing edge some beans swirled out of the hopper. I am not exactly sure what saved our lives, but I believe the tornado lifted up, pulling its punch. When it was over, my father's visored cap was thirty feet in the air, bobbing in an updraft, but he was safe.

Today's storm was a minor leaguer, but, lengthening maddeningly, keeping my father under siege on the forty, it reminded me that—for all the preordered, rehearsed feel to my days on the farm, for all the boredom—the drama of death was always offstage, one moment away from materializing. Once, while horsing around in the haymow, Harvey tied a sling rope around his neck and stepped off a bale. "This is how they used to hang rustlers," he said to Roy and Dale, who were young kids. They watched Harvey's long, urgent kicks aimed at the bale until they realized he couldn't get back on. His face was purpling. His hands went to the rope over his head, but he couldn't lift himself. Roy and Dale seized him then and felt the lead in his legs. Working against panic, sweat pouring off, Roy got Harvey's weight onto his own shoulders, and Dale began to maneuver bales under Harvey, building a platform. Finally it was high enough to take the weight from his neck and loosen the rope. A pail of water was hauled up the mow ladder and splashed on his face. The first thing Harvey said when he came to was, "Don't you guys tell Daddy. Promise you won't tell." The rope had burned a red cinch mark on Harvey's throat. He buttoned his summer shirt to the top, but during the evening milking my father found out. "Jeez-oh-mighty, how many times do I have to tell you not to play around like that!" This was my father on the verge of grabbing a whipping stick, yelling, and surrounded by other yelling from cows that picked up on the agitation and slammed about in their stanchions. My father's anger had a peculiar resonance, which may have come from that day, New Year's Eve, 1934 when Herman Kohn hanged himself in a haymow. Herman had had a falling-out with Heinrich and had bought eighty acres for himself two miles south on Carter Road, where his debts got the better of him. His body was found during morning milking. My father almost never talked about it, but I knew he carried it with him.

Now the storm outside was slacking off. I heard father's pickup in the driveway. My mother and I jumped to let him in the door. He seemed neither flattered nor put off by the relief in our greeting. "I'm starved," he said.

....

At Ittner's in the afternoon my father ordered bags of bonemeal and mineral supplement to mix into chicken feed. Jean affably asked about the chickens, and he said, "I didn't think they'd ever start laying again. Not after this winter. But they're up to fifty, sixty eggs a day, with big, dark yolks. I've been feeding them grass cuttings, which they love."

"You going to sell them with the farm?"

"Not according to present plans." My father handed her a twenty-dollar bill, winking. "Ma needs something to keep her busy."

So the flock of chickens would remain with the garden and the house. His purchase of feed mixes was not merely going through the motions. In the snows of next winter he would be shoveling a path to the chicken coop. There was risk in that, but, all in all, I thought it was a sound idea. That evening, when Diana and I went over to Jean and Tom's house for a supper of fried chicken and nine-vegetable salad, Jean said the same thing. "Your dad seems to be striking the right balance. He's going to keep himself occupied."

And Tom said, passing around beers, "Sounds to me like he's going to be busier than ever. This plan of his to travel around the country is something else." Tom tipped a bottle and swallowed mightily, then stopped. He saw I was puzzled. "Hasn't he told you?"

"I'm not sure," I said.

"He and your ma are going to spend every spring and fall on

the road, traveling around, seeing everybody in the family who's moved away. He told me he's looking forward to Sandra moving so they'll have one more place to visit. They'll be heading your way this fall."

Jean interrupted. "It's supposed to be a surprise, I think. But your dad's so excited about it he's been telling everyone. He and your mom are planning to go to Washington as soon as you have your baby. And then they might circle down to Florida and over to Texas and who knows where."

So my father and mother had plans for their retirement! Grand plans! Calculated plans! They were going to hit the road for extended periods in the pickup and visit their children and an assortment of friends and relatives.

"I think it'll do them a world of good," Diana said.

"They're really looking forward to it," Jean said, "Ever since they saw you guys out in California and had such a good time I think this has been in the back of your dad's mind. To get out and see more."

To see the U.S. of A.? Was that part of the charm, a fulfilling of some held-in, held-off fantasy? I shook my head, wondering.

"He and your ma are going to give you a run for your money," Tom said. "They might put on more miles than you two next year." Diana and I had long been known as "the traveling Kohns," the salutation that at times had been written on letters needing two or three address corrections to catch up to us.

"We're probably going to be out of the running altogether," I said, laughing nervously. I patted the bulge at Diana's waist. "We'll be housebound."

Jean could not let that pass. "Oh, no! You're not going to grow up just because you're going to be a daddy!"

"Not if I can help it," I said, laughing even more nervously.

The last lingering pink was dropping out of the sky. In Michigan, summer nights do not start until about ten o'clock. On five separate occasions in the 1960s daylight savings was the subject of a vote, either by the state legislature or by the electorate, and each vote reversed the previous one. Early-to-rise farmers entered into the debate on the side of morning light, but in the final vote they were outnumbered. I thought of myself on the highways, trying to compensate for late starts, driving into the long-lit evenings, and I thought of my father, up at three-thirty and on the road by four, in the dark of the morning. And I thought of him navigating the countryside, pulling into our driveway, eating at our table, cradling his new grandchild in his hands, and then driving off to another spot on the map, he and my mother taking in the sights, as if their lives had just begun and seventy years on the farm was but a preamble to some absorbing, triumphant travelogue.

At midnight, leaving Jean and Tom's house, our way was blazed by a full moon. In this startling illumination the scarred, unrecuperated fields had the look of winter. At the farm the chickens, tricked by the light, were awake and scratching in the grass cuttings my father had thrown them. They were like jungle creatures furiously active after sleeping away the day. "I guess Roy or Ronald will take care of them," I said, "when the folks are out touring."

....

In bed, Diana said sympathetically that I had lived my life backward, that as a man I had grown up very slowly because, perhaps, as a boy I had grown up too fast. "You told me something, when we were first getting to know each other, that I still find incredible," she said. "That morning your dad first took you out to hoe. How old were you?"

"Six years old," I said. First grade was over, and summer vacation was about to begin. I went outside after breakfast to a pile of sand in the backyard. I lined up my toys. The wild dill was fragrant. The sun inched a notch higher. I heard my father in the smithy, whetting and filing the hoes to a straight-edged fineness, and then he approached and handed me one. "This is yours." I went with him into the fields and came back that evening full of manhood. Six years old! From then on, every summer day was a workday, Sundays and the Fourth of July excepted. It may not be too much to say, as Diana does, that I was not a boy again until my crazy teenage days of carousing and hitchhiking around the country, days that, while tamed down, had persisted in general outline through my twenties and thirties.

"There comes a time, though," Diana said gently. "Don't you think?" Premature gray flecked her hair, and her face looked more settled, more authoritative than on our wedding day ten years before. She took my hand and held it gently to her stomach, alive and thrumming. "I have an appointment with my obstetrician at the end of next week," she said.

"I hear you." I said. "I've been thinking that I should talk to Reverend Westphal."

"About your father."

"Yes—and also about baptizing our baby."

"Our baby is going to be born seven hundred miles from here." I could feel Diana stiffen.

"I'd like to come here at Christmas. For a visit, a holiday. We could have the baptism then."

"Who are you doing this for?"

"I'm not sure I know. For everyone. For myself. For us. I guess I want to make some kind of public declaration or promise."

Diana propped herself up on an elbow, staring.

"I am trying to say I agree that the time has come," I said. "How can I explain it? I have not stood up in a church and identified myself and made a promise about anything, not since I was a kid. Not even when we got married. And—I don't know—I'd like to promise to be a good father, I guess."

"I don't think," Diana said, "that I'll ever really understand the German mind, but I love you very much."

Reverend Westphal, when we asked him about the baptism the next afternoon at his office, said that the Lord would be pleased to receive our baby. Our request, open to skepticism, was far from automatic in the Missouri Synod. But Reverend Westphal had always acted toward me, fallen from grace, as toward a church member in good standing. "I can keep your baby's membership here, and when you finally find a church to join, you can have it transferred," he said. "I know your folks are counting the days till the baby's born. Your dad's all set to go. He's been telling me he's got his maps, he's got the directions. Soon as that baby arrives, they're off."

"I guess you know they've got a whole itinerary planned out."

"Yes, your dad told me." Reverend Westphal smiled. "It's always safe for people to tell the minister their plans. If someone changes his mind, there are no hard feelings. The same with your dad's plans for the farm. He talked to me a little before he made up his mind, kicking things back and forth."

"I wanted to ask you about that. Is he all right, do you think?"

"What does Sandra say?"

"She says he's accepted it, that he knows it's for the best."

"She's right, I believe." He forced me to attention with a look. "Your dad is a remarkable man, and I don't say that casually. He knows what's right and what's wrong, and he's not afraid to tell you, but, by the same token, he is one of the least judgmental men I know. If someone is living with someone without being married—yes, we have people like that around here, just like everywhere—he may wish it was different but he doesn't let his actions show that. With the farm it's been the same way, from what I know. If it was possible for one of you to take it on, that would have been great. But it's not possible. It's not financially practical. And, besides, you've all made other lives for yourselves, and that's just as great. It might even be better."

....

That evening, my father and I sat in lawn chairs under the big maple. "Feels good to sit," he said. The sun was an hour from the horizon. We watched a swallow dart in and out of her nest mudded to an inside facing board on the porch. "I see you put the eaves up," he said. "Thanks."

"Sure."

He stretched out his legs and leaned back. An argument could be made, I realized, that this very Germanic man had not been a very Germanic father, if, as Reverend Westphal had said, he was satisfied that from six children—five sons!—not one was a true Kohn, an heir. Had my father all along, from when we were kids, known he should let us go off and discover ourselves? Could we say that we had left? Or had he let us go?

On Heinrich's deathbed, at home, at age eighty-one, his admonition to the grandchildren brought to see him was, "Remember to honor thy father and mother." Johann uttered the same words as he fought for breath after his final heart attack. The German Lutheran creed was that order descended from God and, through the hierarchy of Creation, was handed down to fathers and mothers, his earthly caretakers. A child's place, on the condition of ultimate obedience and respect, was as the ultimate inheritor. In Milton's reconstruction of the loss of Paradise, Eve was confronted by something beyond obedience and inheritance, the temptation of choice: ". . . of this Tree we may not taste nor touch; God so commanded and left that Command Sole Daughter of his voice; the rest, we live Law to ourselves, our Reason is our Law" —and the temptation, the choice, the Fall gave us the power to be ourselves, to operate outside the order of Creation. In the Garden of Eden, Adam and Eve were not free. German Lutheran children, staying within obedience, are not free.

My father had grown up watching Heinrich and his sons, and he had seen how a German patriarch's expectations can soar and how his sons must reach. Heinrich had five sons. Four lived into adulthood, Wilhelm having died while a baby. Charley, the oldest, was expected to take over the back forty, a homeplace for body and soul, a physical and spiritual inheritance, but one requiring Charley to be a farmer. Charley became like an engine that coughs and knocks and then cuts out when too much is asked for it. A gunnysack on his back, he hitchhiked from odd job to odd job, from beer garden to beer garden, never lasting long anywhere. ("He had one job over by Mount Pleasant, and one morning he was gone, just flagged down a freight train and hopped on. Another time he worked at the sugar-beet factory but didn't like his boss, so he said "Kiss my ass" and quit. He'd come back to the farm to try to make a go of it. He planted chicory one year, and potatoes. The teenagers liked to hang around his cabin because he was like a kid himself, and he always had a jug.") When Heinrich heard that Charley was drunkenly, defiantly offering pieces of land for an evening of drinks in Willard, Heinrich put Johann's name on the deed to the back forty. Johann farmed it for Charley, and they split the profits fifty-fifty. There is no way to know how deeply scarred Heinrich was by his firstborn. Harmony on the farm, I know, was diminished. Herman, the second son, fell into aggressive patterns. "Big shot," Heinrich called him. There wasn't room on the farm for both of them, but to have a separate place Herman had to take out a bank loan, for which Heinrich refused to cosign. Money was easier at the Federal Loan Bank, part of a government program, and Herman borrowed eight thousand dollars. Against his better judgment, Heinrich and a brother-in-law made up the rest, thirteen hundred dollars, and Herman made his move, which ended on a rope after three straight years of losses. ("Even after Herman left, he and his dad didn't get along. Every morning, when his dad hauled milk to the Willard cheese factory with his horses, he'd go right by Herman's place, so he'd stop in, and the two of them would get to arguing.") By the time of Herman's death, Henry, the baby and namesake, was in Bay City, and Johann was a solitary survivor, carrying on.

My father had none of his five sons left, but then you could say he had every one of us.

....

My father was up early, before the worst of the heat, out on the gray Ferguson, cultivating, propped under a sun umbrella. His eyes were down, trained on the cultivator guide, a simple metal stick lined up with the row second on the right. He craned his neck. He saw a pastiche of beans, weeds, cultivator blades, tumbling soil. I followed with a hoe, chopping the weeds the cultivator missed. To understand about hoeing, and the lengths we used to go to avoid it, you have to hear Sandra's story of the afternoon my father said that she could quit the fields early if she cooked a chicken for supper. (First, I had to catch one and chop its head off. I was maybe twelve, and I'd never killed a chicken. I grabbed an ax and chased this rooster round and round and when I finally got him it took me a dozen whacks to get his head off. But I didn't care. I was just so glad I didn't have to be out hoeing.'')

Today was another scorcher, the sky splashed yellow and blue, a tie-dyed look. At noon, my father thought that I, out without a straw hat, had gotten too much sun and insisted I stay in the shade for an hour or so. About one-thirty I walked back to the field, bringing Diana along with a thermos of water for my father. We crossed the creek by the cement bridge. The rainstorm had rejuvenated the current, and a branch of willow from upstream that a muskrat had chewed on went sailing smartly by. The creek was like the reverse of a California irrigation canal, draining the immediate watershed and lunging together with other creeks into the Kawkawlin River, an opaque brown riot, on into Saginaw Bay and east to the Atlantic through the St. Lawrence Seaway. Were it not for creeks, Saginaw Valley farmers would have exceedingly mixed emotions every time it rained. The water at our feet suddenly appealed to me. With shoes off, I waded in. I could feel fingerlings, carp or suckers, investigate my legs. Grassy, relaxed banks sloped above me. Cool as the water was, the larger creek felt warm and intimate and made me feel at home. If ever there is a place to have fun, it is a creek, and I had had fun here, catching tadpoles in a jar with Johanna, spear-fishing on moonlit nights with Johann, astonishing fun, come to think of it, so carefree and transforming.

I heard my father shut off the tractor engine. ''You lost?'' he shouted. We hurried the thermos to him. He took it and drank with his eyes open. A monarch flickered by. My father paused for a breath. A hand scratched violently high on his backbone. Another tug on the thermos, and his cap came off. Sweat had plastered his hair to his forehead. A red bandanna came out. Behind the noncommittal face, behind the unshaved, sun-savaged cheeks, the dripping chin, the impassive home-on-the-range look, was total exhaustion. A final swallow: the dying coolness of the drink.

"You ought to have let Diana stay at home," he said to me. "She'll get heat stroke out here." The return from my high spririts to that chastening remark was another return to my boyhood.

But I had not acted responsibly. The sun was beating on her, and she did not look well. I helped her to sit under a tree, her back against its trunk. "It's just a little wooziness, nothing serious," she protested. "You go and hoe. I can walk back to the house."

At supper that night, to my father's inquiries about her health, she said, all dignity, "I didn't know that pregnant women got special treatment on a farm. I thought we were supposed to be out in the fields until the last minute."

"Times have changed," he said.

All that evening I was aware that my father's conversation was in an idiom of names, his children's names and his grandchildren's names. He filled us in on all the family news that we, arriving late, had missed on our first evening, and he gave us a newspaper headline he had saved, referring to a royal pregnancy in England. The headline said, LADY DIANA AWAITS FIRST HEIR. He kidded us. "I guess when you two finally got around to having a baby it's front-page news."

Yes, times had changed, I thought, and the biggest change, above and beyond everything, was the dream that kept my father alive. It was not Heinrich's dream, not the preservation of peasant Germany, not the preservation of the order of Creation, not the preservation of the family farm, but the preservation, the well-being, the extension of the family. Or perhaps I am wrong,

I thought. Perhaps that was always my father's dream and I didn't have the wisdom to see. No matter. Now I did see. I had advanced far enough into my father's skin to see.

I remembered a night when I was seven or eight. I had dawdled over supper and was five or ten minutes behind schedule in going to the barn, where my father already sat with his milk pail under the warm cows. A snowstorm was blowing. Leaving the house, suddenly the whole world was white and bone-deep cold. The gray mass of barn was gone, and, turning back, the house was gone. The trail between house and barn that we had shoveled in the afternoon had blown shut. I walked in circles, blindly feeling for the cow-yard fence, the chicken-yard fence, the garage, a corn crib, the diesel tank, the watering trough—all gone. I knew I would freeze to death. "You're going the wrong way," my father yelled in my ear and caught me in a man's rough, here-boy, saving grab. In the kitchen my mother ran cold water into the chipped white washbasin. My father plunged my fingers into it. It was unbearable; the cold water was hot as dry ice. My father took out my hands and placed them in his. The agony of my frostbite seemed to extend from my hands into his, smelling freshly of the cow barn, and up his forearms to his tightened face, normally so easy to behold, and back to his hands, which looked like they were squeezing the life out of mine but which felt soft, and which, on the principle of sympathetic magic, were removing my pain and taking it up into him like so much of everything else he had taken in and held inside over the years.

....

The following week, in time for Diana to make her doctor's appointment, she and I went home to Washington. We told everyone we would be back at Christmas. Before we left there was a baby shower, arranged by Sandra and my mother. Aunts, cousins, friends, neighbors, those who would have been at our wedding if Diana and I had not been married in a house in San Francisco, came now to a baby shower. The old farmhouse shook with card games and door-prize drawings and gossip and much celebration. Diana's face was flushed. She thanked everyone over and over. In the kitchen, keeping out of the way, my father and I listened to the Tigers on WSGW.

On the ride back east Diana asked me to suggest names for our baby. Did I want to name a son "Fredrick" is what she meant. At the baby shower Sandra had pointed out that, after nine grandchildren, my father's name was still available, consciously unused. It had been left for me as a means of Germanic redemption.

"I think we can choose any name we want," I said. "One name is as good as another."

"Are you sure?" Diana asked.

"Yes."

I let out a breath and reached a hand over to Diana's stomach. The farm, up for sale, was at the other end of the road we were traveling. I had thought the farm had a voice, crying out in echoes audible only to a Kohn. But it must be that a modern heart, so empty of sentimentality and faith, is suggestible and aches to hear. It hears what a writer might call literary possibilities. A writer, a creature and a captive of the third person, cannot finally, I had decided, do anything but write, as a farmer must farm. It was not until later—in fact, not until I sat at the typewriter and wrote this book—that I fully realized how much my writing is like my father's farming and how much I had become like my father. Writing and farming are endeavors of the solitary, driven soul, driving against the odds. A space is held out for accomplishing something, for answering the hope of individual worth. You are made to take control of your life, as best you can. Writing and farming enlarge also the chance of individual consequences. You can fail. Success in one year is no guarantee of success in the next. The winds of change blow against you. If writing and farming are in decline these days, it may be because of the lost capacity to go it alone, to take risks, and to be true to yourself.

At the baby shower, in a moment of reflection, my father had said, ''It's for the best, Sandra going off. She and Mike will be able to work hard and make a good life for themselves. Like you did.'' I believed I knew what he was saying to me, and I believed we both had come to understand that escapes and estrangements and long silences are sometimes only love out of proportion.

In the car, the green summer whizzed by.

EPILOGUE

I guess when you give up things something good can come of it. My blood pressure is way down.

—from a letter, 1985

My father had set a schedule for himself, and he kept to it. In December he signed a land contract, a joint venture with Earl Gerstacker and Vern Chapman, and the Kohn farm, save for the land and buildings in the immediate orbit of the house, passed into their hands. The two new owners were from the Korean War generation. They were neighbors, fellow parishioners, lifelong farmers and of German pioneer stock, men compatible with my father. Don made a bid that my father turned down, which might have led again to hard feelings, but the two men were now past that. Don had shown forbearance, allowing my father to have the last word, as it were, and my father had said with a nod of empathy, ''Good luck. A farmer today has a tough row to hoe.'' The fact was also that Don had lost some of his enthusiasm for big-time farming. A *Bay City Times* reporter had interviewed him, as one of the valley's success stories, for an article about the 1980s farm crisis. ''You're always looking for that year when you're going to hit,'' Don was quoted. ''But it's not there. If things don't change, you're going to see less farming.'' The headline over the article said: FARMERS NOT SURE SONS SHOULD BE TOO. ''Don Rueger is a third-generation farmer in the Auburn area. But he says he doesn't know if he should encourage his fifteen-year-old son to follow in his footsteps,'' the article began, and it ended with this statement by Don: ''To be a farmer today, you really have to like what you're doing.'' My father had mailed a copy of the article to me, along with one about the farm crisis from *Newsweek.* ''I keep reading that farmers aren't supposed to be farmers anymore—they're supposed to be businessmen,'' he wrote. He and Don were making the same point: that the only real satisfaction in being a farmer comes from the magical-seeming power you have to define yourself and your work. As my father put it, ''If I'd wanted to be a businessman, I'd have gotten a job with Dow.'' He wrote that he was ''glad to be done with farming, the way it is.''

For a long period I read his letters closely. Could I find any sign of bitterness? Of a hollowing regret? I found none. I found instead a steady release of funny little stories, of the type he had started to tell the winter before, when things were going bad. He wrote that several parishioners were unhappy with the taste of the Communion wine, bought from a local winery in solidarity with a ''Buy-Michigan'' campaign. ''It's the awfulest-tasting stuff,'' Mrs. Westphal had said to him, and he had joshed her: ''That's so you only take a sip, not a whole slug.'' This joshing, wisecracking guy had emerged, it seemed, from the cocoon of the farm. But when I mentioned this to Jean and Tom over the Christmas holiday they said, putting me straight again, that the lighter side of Fred Kohn was not new to them; they had seen it many times. It was new only inside the family, and it was new especially to me. I had not noticed it before, even when I might have, because it was not part of the man I thought I knew. I had changed and become more observant. I saw my father more complexly. I saw he was not always a hard man, but a man of dimension, and when I reread some of his letters from years ago, here and there were the same kind of funny little stories that had seemed so new. Yet, I thought, he had changed, too; it was not my imagination. He had softened. At a church picnic that Diana and I had attended the past summer my father and mother were conspicuous by their presence. The picnic was a toned-down version of the old hoot-and-holler July Fourth affairs, and the profit was to go to the centennial project. The turnout was a disappointment. Many of the younger families were off swimming or boating, and among the old-timers my father and mother were about the only ones who did not leave after the noon meal. They sat with Reverend and Mrs. Westphal at a wooden picnic table and watched the centennial-committee members try to drum up interest in the old carnival games. Roy and Lorie were in charge of the ring toss. While the kids tried their luck, my father clapped encouragingly. This did not seem like the man who had been so contrary to the centennial committee, and, indeed, later, with money from the sale of the farm, he and my mother contributed a thousand dollars to the project. The gift was made in honor of their wedding anniversary and was announced in the weekly church bulletin.

Whether others in the congregation took it as a signal I can't say, but a consensus was building. The centennial committee had compromised on a less elaborate floor plan and a budget of about four hundred thousand dollars, nearly half of which was now donated or pledged. A work crew, after laying a cement foundation for the one new wing, had paused for the winter. The fresh cement was under snow when Diana and I, returning joyously, kept the appointment for our baby's christening. Reverend Westphal held a private evening service for us. He was in the moonlight shoveling a heraldic path as we pulled up. At the baptismal font, he sprinkled water and said the sacramental words, "I baptize thee, Jennifer Suanna Kohn . . . " For a while Diana and I had considered naming our baby Johanna, then had decided it was unfair for a twenty-first-century woman to have a nineteenth-century name, but even as we made that decision we began to remark on Jennifer's good spirits, so much like

Johanna's. Jennifer was born with them, and as she grows, my daughter reminds me ever more vividly of my grandmother. Reincarnation is too mystical a concept for German Lutherans, but they do believe, from Ecclesiastes, in natural cycles that double back on themselves, and in a time without end that follows the curve of infinity. The past catches the present; one generation is visited upon the next. It is the German Lutheran vision of immortality. For myself, I get a warm feeling every time Jennifer smiles, and it is warmer by a few degrees because I can hope that her life will hold some of the satisfactions that Johanna's did.

One way or another, for better or worse, we all have the past with us, and it may be that mine will always be for me a paradox of ancestors I never knew, and a father I got to know so late, and another ghost—my own—that I am still tripping over. Isaac Asimov got it right, I think, in his short story about a time traveler who is scared by a stranger making noises in the room next to his. But shouldn't the traveler have guessed that he himself is the stranger carrying on life in a former time? And shouldn't I have known that a son, when he comes home, no matter the kind of home, no matter his age, cannot help but become in some fashion the boy he used to be? Don't all of us have to complete a circle of comings and goings before we grow up?

My father left the farm for the war, a big, handsome fast-learning farm boy with wavy, dark hair and an eye for the ladies, and while thousands of other farm boys found better opportunities after the war in California or Florida or wherever, he returned home to rural Michigan. For the next thirty years he milked his cows by hand twice a day every day and did not miss, in total, more than a few days, until that incredible time in 1974 when he and my mother came cross-country to meet my new wife, riding a Greyhound bus to Nebraska, switching here because of a drivers' strike to a Trailways bus, and showing up after three days and three nights on the road at our communal house in San Francisco. On that trip they were gone from the farm almost three weeks. Afterward he bragged about it for months, as I found out much later from Sandra and Jean and Reverend Westphal. "It was such an adventure, going all the way out there to see you. He couldn't stop talking about it," Sandra said, telling me this at the baptism, "I agree with what Jean says. I think that trip was the beginning for him. Here was something he could do that he enjoyed and was just as important as the farm. He could go see you and Harvey and Dale; he didn't have to wait for you guys to come here for a visit. I think that's when he started to let go of the farm, though none of us realized it at the time."

The day after Diana came home from the maternity ward with Jennifer, ten years after that California trip, my father's pickup had roared into our driveway in suburban Washington. He and my mother had left the farm in a hurry, before dawn. He had not shaved. Black whiskers of a held-over virility darkened his jaw. He unloaded jars of jams and pickles and sauerkraut. He unloaded a pine dresser that they had saved upstairs from when Liz was a baby, and an oak dresser that had been mine as a boy. My mother handed me a potted shamrock and envelopes of garden seeds. There also was a paper bag in which were some of my old toys. In the following days, while my mother and Diana's mother cooked meals and managed the house, my father split a cord of wood for our fireplace, dug postholes for a wooden fence, built bookshelves, raked leaves, and generally made himself handy. Some days I worked alongside; other days I was at my typewriter. Evenings we took turns holding Jennifer. The farm was not completely out of mind. He had combined his windblasted navy beans ("Got about thirty bushels to the acre, not bad, considering"), and he had a field of corn left to harvest. Every morning he checked our copy of the *Washington Post* for reports on Michigan's weather ("We need a good long dry spell"). Once he phoned back to Michigan to ask about the chickens ("They're laying good").

On the day he left in the pickup, I thanked him for all the work he had put into our place. "Hate to see you go," I said. "I've got a few other jobs I could use your help on."

"Save them," he said. "We'll be back."

....

Six months later, in April, my father held an auction to get rid of things no longer useful, and it was my turn to help out one last time.

Diana and I, with Jennifer in her car seat, drove to Michigan. I found myself liking the intermediate sense of drawing near the place, seeing the Germanic homesteads and the green and brown fields. The sky was a meringue of clouds, white, showy, harmless. We arrived in time for supper, and after we ate my father held forth with the news. A hole had been blasted into the east side of the church, making Sunday services drafty, but keeping the centennial project on schedule. Oscar Ittner had undergone more tests and was about to have triple bypass surgery (twice, as it would happen, but with a remarkable recovery, and with an accompanying *Bay City Times* feature photograph of him working out on the therapy machines). The faith in the future that the Ittner family had shown was beginning to pay off. Farm prices had stabilized, and the valley farmers could see their way past the worst of these hard times. In Bay City, money had been found to build a replacement for the collapsed Third Street bridge. In Midland, nine months after construction had halted at the nuclear plant, Dow Chemical and Consumers Power were mired in a legal feud of suit and countersuit (which they would eventually resolve) and the centerpiece of it was to be a new partnership under which the unfinished nuclear plant would be converted into a plant fired by natural gas. Just the talk had helped set off a new boom in the valley. Energy wildcatters were drilling for natural gas on many of the same farms that had been drilled for oil. The oil boom, for the time being, was over in the valley, as it was everywhere, because the world marketplace was suddenly awash in petroleum. But that, in turn, meant lower fuel bills for diesel-driving farmers. All in all, my father said, the mood in the valley was that "things are looking up."

As for my father's mood, after an upbeat evening with us, he was not exactly fired up the next morning for our task, which was to comb through the sheds and the barn in preparation for the auction, but even in this final act—a time certain for second thoughts—he was not fazed. He was as determined as ever, and his very equanimity made me think that my past worries about his will to live must have been farfetched and paranoid. He had come out of a heavyhearted year without a heavy heart.

Going from building to building, he indicated to Ronald and Roy and myself the things to be auctioned off: a set speech, I realized. He had already walked through here and made his decisions. Only the gray Ferguson, one plow, one cultivator, his best hoes, his best axes—only the few pieces of equipment relevant to his reduced, one-acre farm—were to be kept. He would not sentimentalize the past. The four of us began to haul milk cans and rolls of barbed wire into the old cow yard for public viewing. "Leave enough room so folks can get up close and see what they're buying," my father said to us. At first we were a little awkward, faltering a bit, waiting for him to take the lead, but then we got down to it. There was real work to do. We crowbarred the hammer mill off the barn floor, pulling up the long spikes. We unbolted the bean pickup from the Massey-Ferguson combine, taking a minute to shine with a cloth the combine's red chassis and flappers. From a loft in the tractor shed we lifted down old horse-drawn implements from Heinrich and Johann's era that my father had not been able to adapt to his tractors.

Wrestling with a one-horse walking plow, Ronald and I banged our shins. "Damn piece of junk," I said.

"Used to be junk," my father said. "Now it's a collector's item."

The plow went into a lineup of such items in front of the barn, the antiques that would attract dealers. "Don't have to pay income tax on the money I make from these, according to what the auctioneer tells me,'' my father said. ''Anything that's older than 1933, from before the income tax was passed, is tax-free. All the rest, I've got to give the government its cut.''

Among the items that predated 1933 were three one-man corn planters. Heinrich used to march up and down his fields, stabbing a planter into the ground, dropping the kernels out serially, scuffing dirt over the hole, creating rows. Also there were the following: a hammer for working with stone, broadaxes, adzes, crosscut saws, cant hooks for rolling logs, chains for yanking out stumps, dynamite augers for blowing them up, a posthole digger, a rope-and-pulley fence stretcher, a hay knife, hayforks, a forge, dozens of smithy tools, a hog kettle, butchering knives, round scrapers for removing hair from the carcass, U-hooks for hanging it by its hind-leg tendons from an icehouse beam, a sausage stuffer, and the hand pump from the well. These antiques felt usable, and they were. At one time or another, my father had used every one of them.

At the end of the day, he made a final tour of the sheds and the barn, now like echo chambers. ''Good, good,'' he said. ''Good work.'' He took off his cap and walked bareheaded into the house.

''Who would've thought he'd ever sell all this stuff?!''Ronald said, marveling.

''I know,'' I said. ''I never thought he'd be alive to see this. I figured we'd end up doing it after his funeral.''

The following day in a fierce gale, more than a hundred farmers roamed about our yard while the auctioneer praised the wares. The bidding was brisk, and by nightfall the auctioneer's assistant, his son, had a cashier's box full of cash and checks. Even the old stone-and-cement watering trough was sold, although the money for it had to be refunded when it proved to be immovable. Not until the last transaction was complete did my father step a foot outside, as was the protocol, but a number of farmers, wiping mud from their boots, came into the house to shake his hand and wish him luck. For many, that was the sole reason for their appearance at the auction, and they left empty-handed. Others were keen to buy. The young farmer who had been by two years before to ask about the combine won the bidding for it. Another farmer, a middle-aged man, said to me, "Tell your dad I bought his hay baler, and it'll be put to good use."

In the morning, surveying the scene outdoors, my father complained mildly about the car ruts in the lawn. He stood with my mother next to a patch of strawberries that was showing life. "Ought to get a good crop of berries this year," he said. He scouted through the garden, looking for winterkill. My mother snipped off dead branch tips from a peach tree. She looked forlornly at hundreds of tiny teeth marks near the base of the trunk. Mice in secret tunnels under the snow had killed a pair of peach trees. "We can plant some more. They grow fast," my father said. His optimism sounded natural enough, but all the same, on this particular day, I think he was glad that most of his children and grandchildren were hanging around and that the conversation, self-consciously on our part, did not dwell on either the past or future.

Other topics were not hard to come by. Important changes were happening in our lives, too. Ronald had lost the job he had held since college at Helfrecht's machine shop, which had gone bankrupt due to overzealous modernizing and expansionism, but Ronald, with an excellent reputation in his trade, had landed a new job, superior to his old one, in the mechanical end of research and development at Dow Chemical, and was "set for life." It had unplugged energies in him. He was attending special courses to upgrade his skills and already was moving up in rank. On certain days he would pass within hailing distance of Roy, who, after his layoff from the nuclear plant, had thought about pulling up stakes and moving west or south, but who, with a hardworking record to recommend him, instead had been grabbed up by Stedman's Construction, which was prospering on a regular run of subcontracts from Dow Chemical. No doubt Sandra also could have found a job in the reviving valley, but she was bursting with excitement for York, Pennsylvania, where four years of schooling and pennypinching had paid off for Mike in a cartographer's position. His first big assignment was to map out the American military bases in Germany. Sandra, having said good-bye to friends, neighbors, co-workers, kindred souls of the church, softball teammates (the reigning class C champions of Midland!) and everyone else of meaning to her in the valley, was confident of finding replacements in York, which is not to say she would not shed a few tears along the way. As for me, I was about to become reacquainted with my first daughter, Liz, almost eighteen, who, with substantial conflicts and misgivings, had decided to test herself, and me, by enrolling in the University of Maryland, five miles from where Diana and I lived. Liz planned to spend summers at our house. Over the phone I had become "Daddy" to her again, after several years of being "Howard," which I took for something like forgiveness.

In the summer of 1986, finished with high school, Liz came east, and my father and mother moved her belongings in their pickup—furniture and record albums and clothes and a framed poster of Jane Fonda kissing Kris Kristofferson from a movie I had helped write. I had forgotten that I had given Liz the poster. "I saved everything you sent me," Liz said softly. And so we began a life together, which I can say, two years later, has been a good time for us.

A man with two daughters, one old enough to make him overnight a grandfather, must be grown up. Diana and I are in our sixth year in the same house, and, in cutting back my travels by about half, I have gotten to know neighbors and have made friends. It is not unusual for me to be out at dawn in our backyard that six years ago was half an acre of vacant lot and is now an enviable garden. I have a sense of roots where we live. I have put aside many boyish things to be a father, although, to tell the truth, when I am weeding my garden I sometimes do get the feeling of time warping around me—of the man and boy walking in parallel. I use an old hoe, worn to two inches of flat blade from an original five. My father gave it to me the day after the auction. He had kept it for me. "Might be of some use to you," he said, eying it before he handed it over. I could not get out a response from a well of emotions. I touched the blade. He had sharpened it with a file. Of course, I could now buy a new heavy-duty hoe at a hardware store, but I like working with this old one, and, without quite realizing it—Diana caught me at it and pointed it out—I have begun to teach Jennifer the names and leaf patterns of weeds.

Our garden owes some of its thriving look to my father and mother, who have continued unstintingly with their visits. Their pickup is wheeled into our driveway, and they get out and go to work: it is their new routine. My father has since told me that, as a young man, he had written across his cardboard suitcase, CALIFORNIA OR BUST, but, between the Depression and World War II, there was never an opportunity to pack the suitcase and hit the road—"So I had to wait all these years to get it out of my system."

A friend of ours, who had known my father as a farmer, asked him how he spends the time he is not traveling. "Oh, we've got chickens and our garden," he said. "That's more than enough for Ma and me."

And it is enough, I realize now, more than enough, when you have kept your sense of yourself. Character is all that any of us have at the end, the sole property that is ours. I don't mean to rationalize. My father loved his farm, but he understood better than I the ironies implicit in passing on a farm in your own image. The land mocks the farmer by outlasting him and outlasting his family, no matter the number of successive generations. The one thing of permanence that a farmer can bequeath—a life of respect and respectful virtues—will be rendered ironic and pathetic if he begins to act as if he is entitled to a bailout, whether from the government or from his children. My father had to work at understanding this. It was not given. It was an achievement, like any work. I had thought of Heinrich as a pioneer, going off to a new land, and I had thought of my father as a stand-pat guy. But it was my father who had geared himself up for the bold stroke, who saw that the farm did not hold us together, as I had thought, but stood between him and his children. So he had sold it and brought us back together, or rather had gone off to find us, all of us in our own places.

ABOUT THE AUTHOR

Howard Kohn is a former senior editor at *Rolling Stone* and the author of the acclaimed *Who Killed Karen Silkwood?* An award-winning journalist, his work has appeared in *The New York Times, Esquire, Mother Jones,* and many other leading periodicals. He lives in Takoma Park, Maryland.